SAMURAI
WEAPONS & FIGHTING TECHNIQUES

SAMURAI
WEAPONS & FIGHTING TECHNIQUES

THOMAS D. CONLAN

amber
BOOKS

This Amber edition first published in 2022

Copyright © 2008 Amber Books Ltd

Published by Amber Books Ltd
United House
London N7 9DP
United Kingdom
www.amberbooks.co.uk
Instagram: amberbooksltd
Pinterest: Amberbooksltd
Twitter: @amberbooks

All rights reserved. With the exception of quoting brief passages for the purpose of review no part of this publication may be reproduced without prior written permission from the publisher. The information in this book is true and complete to the best of our knowledge. All recommendations are made without any guarantee on the part of the author or publisher, who also disclaim any liability incurred in connection with the use of this data or specific details.

ISBN: 978-1-83886-214-5

Project Editor: Michael Spilling
Designer: Joe Conneally
Picture Researcher: Natascha Spargo
Illustrations: Wes Brown

Printed in China

CONTENTS

Introduction	6
MOUNTED SAMURAI	26
THE SKIRMISHERS	58
THE PIKEMEN	84
COMMANDERS	102
FIREARMS	142
CANNON AND ARTILLERY	180
Glossary	210
Bibliography	214
Index	218

INTRODUCTION

> 'We suffered severe casualties, our forces withered to nothing, and we fled.'
> TSUCHIMOCHI NOBUHIDE (1336)
>
> 'The way of the samurai is found in death.'
> YAMAMOTO TSUNETOMO (1717)

THESE TWO STATEMENTS, WRITTEN NEARLY 400 YEARS APART, EPITOMIZE THE CHANGING ATTITUDES OF THE SAMURAI. THE FORMER, WRITTEN BY TSUCHIMOCHI NOBUHIDE IN THE SECOND MONTH OF 1336, SHORTLY AFTER THE ONSET OF CIVIL WAR, REVEALS A PRAGMATISM THAT WAS LOST ON LATER THINKERS, SUCH AS YAMAMOTO TSUNETOMO (1659–1719), WHO TRIED TO JUSTIFY THE RATIONALE FOR SAMURAI EXISTENCE AFTER A CENTURY OF PEACE. NOBUHIDE KNEW THAT IF HE DIED, HIS LANDS WOULD BE FORFEIT, HENCE HIS SURVIVAL MATTERED MORE THAN HIS REPUTATION. BY CONTRAST, TSUNETOMO, WRITING IN A RADICALLY DIFFERENT SOCIAL AND POLITICAL ENVIRONMENT, EMPHASIZED THE IMPORTANCE OF DEATH, FOR A WILLINGNESS TO DIE NOW SERVED TO DISTINGUISH SAMURAI FROM THE REST OF SOCIETY.

A depiction of the Hogen Disturbance of 1156 by an artist of the Tosa school during the Azuchi-Momoyama Period (1568–1600).

Yamamoto Tsunetomo, a samurai warrior served his lord faithfully as an advisor and administrator, and, if need be, followed him to the death. Personal survival mattered less than the continuation of one's lineage, as office, rank and status were all contingent upon the favour of one's lord. Hence, Tsunetomo defined the ethos of a samurai as one who served unto death.

To state that the identity of the samurai had changed is in itself a bit of a misinterpretation, for if Tsuchimochi Nobuhide had been called a samurai, he would have bristled with anger. The word samurai means 'one who serves' and implies subservience to a lord. Tellingly, this term, which today describes the warriors of Japan, and is used in this book accordingly, did not designate the most prominent fighters of Japan's thirteenth and fourteenth centuries. They considered themselves to be honourable housemen (*gokenin*) who would lead a band of followers (*samurai*) in battle. (These *samurai* will be referred to in italics, while the non-italicized 'samurai' will reflect more modern usage and designate Japanese warriors in general). In other words, through the fourteenth century, the word *samurai* referred to followers of a powerful warrior – they either resided in his house, or lived on his lands, but could not fight independently in battle. A *samurai* was to a *gokenin* like a squire to a knight. Imagine, then, how different accounts of medieval European history would be if by the sixteenth century all knights came to be known as squires.

Yet this is precisely what happened in Japan. The descendants of the old warriors, who valued their freedom above all else, gradually lost their autonomy as the cost of war increased and armies expanded from several hundred to several hundred thousand by the late sixteenth century. Ultimately in 1588, Toyotomi Hideyoshi (1537–98) forced all warriors to choose to renounce their lands and receive a guaranteed stipend in exchange, or to keep their lands and pay taxes on them, but to do so as a peasant (*hyakushō*). As an added benefit, all who decided to become samurai were granted the privilege of wearing two swords, a longer *katana* and shorter *wakizashi*, which symbolized their status and allowed them the right, rarely asserted, to

A high quality suit of armour as it was worn in the late thirteenth century. Note the long bow, bearskin boots, swords and the doughnut-shaped tsuruwa, *which was designed to keep bowstrings dry.*

OPPOSITE: *Nineteenth century woodblock print showing Toyotomi Hideyoshi, who rose from obscurity, to become the leader of Japan. He transformed Japan and created a clearly defined samurai status. Here he is portrayed as rousing his soldiers by blowing a conch shell trumpet (*hora*) prior to the battle of Shizugatake in 1583. Hideyoshi lacked charisma, being short and uncannily resembling a monkey (something that the woodblock artist chose to ignore), but he compensated by being a brilliant general.*

INTRODUCTION

WEAPONS AND FIGHTING TECHNIQUES OF THE SAMURAI WARRIOR

Samurai and court nobles were skilled archers. Accuracy was prized, and rosters of the participants of matches, as well as lists of how many targets they hit, survive to this day. Early samurai characterized themselves as adherents of the 'way of the bow and arrow' but court nobles, such as those depicted here, were skilled as well. This illustration is of nobles of the Heian era (794-1185).

cut down any offending commoner who showed them little deference or respect. By contrast, those who decided to be peasants could own their lands, but they did not have the right to own any weapons to defend themselves with.

LANDLESS WARRIORS

Hideyoshi, Japan's de facto leader, a man of obscure origins and great intelligence, managed to navigate the turmoil of Japan's sixteenth century and established himself at the summit of political power. He believed that warriors fighting for their lands caused the endemic warfare of his times and therefore strove to make the abandonment of their lands as appealing as possible by granting those who renounced them membership in a distinct social order of the samurai.

The sacrifice Hideyoshi demanded proved great because many warriors identified with their homelands. Many, in fact, took on the name of where they lived. So strong remained the pull of one's lands, that some, such as Utsunomiya Shigefusa, turned down Hideyoshi's offer for an increased stipend because it would have forced him to abandon his home in northern Kyushu, a place where the Utsunomiya had lived for centuries. Arguing that he could not abandon the graves of his ancestors, Shigefusa sparked a war with Hideyoshi, which pitted him and nearly all the men of his province against Hideyoshi's combined forces drawn from most of Japan. Ultimately, Shigefusa perished, preferring to die with his ancestors than be buried elsewhere. Others acquiesced, however, and became like potted plants that were moved from one locale to another. They never forgot where they came from, and in cases where they got into trouble, even as late as the eighteenth century, they would flee back to their homelands, where the peasants remembered and protected them as fugitives. From the seventeenth century onwards, however, most warriors became salaried officials, and city dwellers, whose position and income depended on upon the favour of their lord.

A great divide exists in Japanese society. Before 1588 the samurai did not constitute a clearly delineated order. Rather, nearly all members of society – peasants, merchants, women, priests and of course samurai – were armed. All who fought could gain a name, and lands, and ultimately great political power as warriors. By contrast, those warriors with great lineage could vanish if for one generation they suffered defeat or ignominy. After 1588, however, all had to decide whether they would become peasants and own land, but by doing so give up their weapons, or whether they would keep their weapons, ensure the continuation of their line.

The creation of a landless warrior elite did not unfold smoothly. Some peasants residing in remote regions hid at times dozens of their swords, and a few prominent warriors decided to keep their lands, rather than the trappings of warrior status, and successfully registered the equivalent of many acres in their names. Some of the most powerful magnates, known as *daimyō*, were never relocated, and this enabled older patterns of lordship still tied to the land to remain. Fundamentally, however, the samurai of the seventeenth, eighteenth and nineteenth centuries were urban dwellers who rarely fought. These samurai became more intellectual and bureaucratic, whose experience with weapons was confined to antiquarian studies or as practitioners of the martial arts. With time on their hands, and no need to defend their lands, samurai intellectuals of the seventeenth and eighteenth centuries, such as Yamamoto Tsunetomo, started to question their role in society. Some studied Confucianism, and perceived the warrior as a benevolent exemplar who governed the country based upon his superior knowledge and virtue. Others practiced the martial arts, which became important only when

peace was re-established in the seventeenth century. And finally, some glorified death as a defining feature of their order precisely because, in a time of peace, sudden and violent death proved elusive. Such attitudes were impossible in earlier ages, for death was all too easy during times of war.

MARTIAL ARTS AND MYSTICISM

The establishment of peace remained a gradual process, for men were accustomed to shedding blood for a generation after the last civil conflicts of

An early photograph showing men wearing samurai armour. The man to the left holds a naginata *or glaive, while the one in the centre has protective face armour, and holds a long bow, while the man to the right wears a conical helmet, popularized in the sixteenth century, and a short pike (*yari*).*

The seventeenth century swordsman Miyamoto Musashi was ambidextrous, and a particularly lethal swordsman. He became a figure of legend over the course of the Tokugawa period, and kabuki *plays dramatize his exploits. Here, even armed with wooden swords, the actor playing Musashi strikes fear into his opponents, who recoil from the prospect of fighting him in a practice dueling hall.*

1615. Until the mid seventeenth century, men remained brutal at heart, fighting in duels, which frequently resulting in maiming or death. Miyamoto Musashi (c.1584–1645), an ambidextrous swordsmen, proved capable of killing many a hapless opponent and wrote the famous *Book of Five Rings*, a meditation that typifies the attitudes of the duelling samurai. Duels were not particularly codified according to school of training or type of weapon, but rather retained some of the brutality of actual combat.

A certain Yagyū Muneyoshi, for example, claimed that he could defeat the best horseman with his sword, and he proved successful by smashing his opponent's steed on the head. Authorities of the Tokugawa *bakufu*, the warrior government that administered Japan from 1603 through 1867, later prohibited members from different schools challenging each other. The wooden sword fell out of favour for one made of bamboo, as the ability to show off agile moves rather than injure or maim and opponent became the essence of martial arts.

Practitioners of kendō *emphasize the form of their moves. Here is a basic guard pose.*

KENJUTSU

All schools of marital arts legitimized themselves through links to the past, but only some weapons were remembered. The battle axe, for example, did not generate its own school, and so has been largely forgotten. Pikes, too, and giant 2.1m (7ft) long swords were also favoured weapons, but did not lend themselves readily to training, and so were largely abandoned. Most surviving giant swords were actually shorted during the Tokugawa period (1603–1867), which accounts for their rarity today. Longbows had become largely obsolete by 1600, but were resurrected as being suitable for a martial art, for firing them with skill required considerable time and training, something that the samurai of the seventeenth and eighteenth century possessed in abundance. Archers competed against themselves rather than slaying opponents, and some accomplished acts worthy of inclusion in the *Guinness Book of World Records*. One archer shot a target 12,780 out of 12,910 times over a period of 24 hours.

The goal of *kendō* was to show skill in moves, rather than brute force in smashing one's opponents. Here, the basic overhead swing posture is illustrated.

KENDŌ – POINTS OF CONTACT

Kendō practitioners went from relying on wooden to bamboo swords, so as not to injure opponents. Points are garnered in matches by lightning strikes to the points of contact shown below. Modern practitioners wear masks to protect the neck, face and head.

MEN

MIGI-MEN

HIDARI-MEN

TSUKI

HIDARI-DŌ

MIGI-DŌ

KOTE

INTRODUCTION

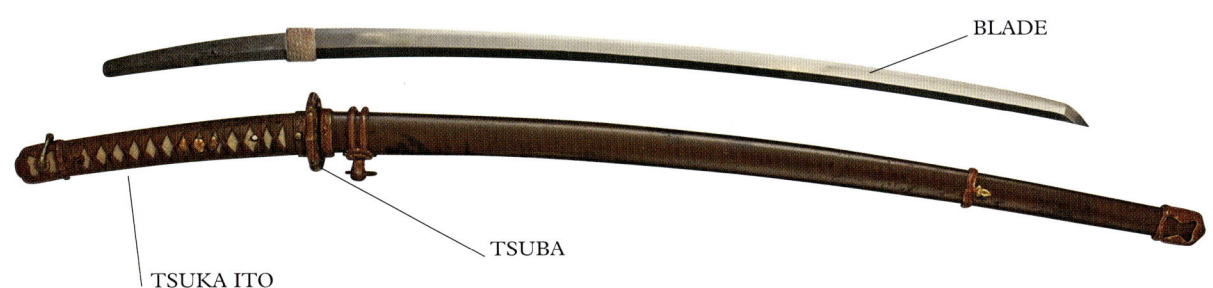

The creation of schools of martial arts influenced memories of the past, as it became assumed that samurai engaged in choreographed and agile sword fighting, while memories of using them to bludgeon opponents was lost.

Most importantly, the *katana* became known as the defining object of the samurai, or in popular parlance, the 'soul of the samurai'. Schools of swordsmanship developed, with great emphasis placed on *kata* – a fixed sequence of moves. Members of one school would not directly compete against those of another, and so again, the sword became endowed with almost mystical qualities. So strong became these myths that even later samurai would come to believe in the paramount importance of the sword. One, named Eto Shinpei (1834–74), a samurai who had translated the Napoleonic codes into Japanese, led a quixotic insurrection in 1875, relying only on swords, but failed miserably. His rebellion has been confused with Saigō Takamori's last great samurai rebellion of 1877. Saigō (1827–77) has been immortalized as the last samurai, and has even been portrayed in recent films (*The Last Samurai*) as upholding an old order and relying only on the bow and swords. In fact, Saigō started his attack by raiding a government armoury, but his trust in munitions has been largely ignored.

Indeed, nearly all have forgotten that his home province, a supposed hotbed of samurai nostalgia and a place where the samurai have been thought (in common memory at least) to have thoroughly rejected use of Western weapons, had a fully operational blast furnace since 1857.

This historical amnesia affected military doctrine of the Japanese Army in the early twentieth century, with disastrous consequences, as *banzai* charges by sword-wielding soldiers caused the unnecessary death of thousands, who were no match for high-

Above: A typical katana *sword dating from the 1930s. The slightly curved blade contains temper lines, for the steel along the edges is harder than that of the blade. A small detachable hand guard, or* tsuba, *protects the hand above the handle, which is made of wood, coated in sharkskin and wrapped in a cloth called a* tsuka ito. *All sword fittings can be removed from the blade.*

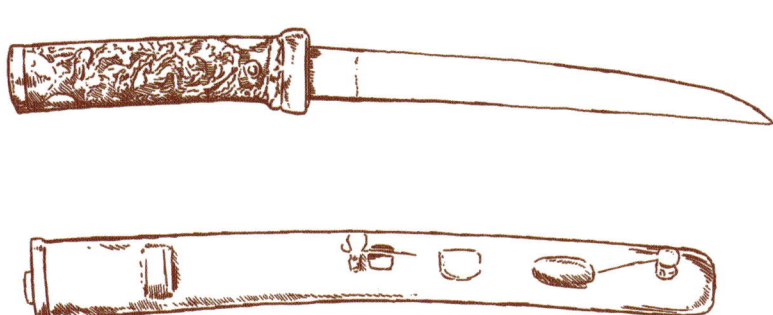

A tantō, *or dagger, with sheath. These daggers, with a blade invariably less than a foot in length, excelled at stabbing and could be easily concealed. The fittings depicted here were replaceable, and in this case, do not include a tsuba or have a handle wrapped in cloth.*

15

SWORDSMANSHIP

KUKISHIN RYŪ KENPŌ

Practitioners of *Kukishin ryū Kenpō*, or the 'Nine Demon God Lineage of Swordsmanship', claims to have its origins in the fourteenth century, and utilizes a much longer sword than is typical for later schools of *kendō*. Although some of these techniques might have existed in the fourteenth century, this emphasis on form, speed and supple movements, is typical of seventeenth-century schools of martial arts. Here several stances are depicted.

KASUMI NO KAMAE

HASSO NO KAMAE

TENCHI NO KAMAE

CHUDAN NO KAMAE

FIGHTING WITH TWO SWORDS

During the early seventeenth century, when masterless samurai engaged in duels to test their prowess, the ambidextrous swordsman Miyamoto Musashi defeated every opponent. Musashi could simultaneously wield two swords, which allowed him to use both to block any attack, and then, while crossed with his opponent's blade, Musashi could extract the *katana* in his left hand and then easily cut down his rival.

powered rifles and artillery. The earlier way of fighting of men such as Tsuchimochi Nobuhide, who fled with his forces rather than suffer extreme casualties, had been forgotten. Instead, the 'way of death' became an operational principle of the Imperial Japanese Army, which did little to preserve the lives of its troops, or for that matter, the lives of prisoners or civilians under occupation.

AN IMAGINED PAST

Emblematic of this reliance of an idealized notion of the samurai, divorced from past practice, army officers fixated on the *katana* sword. Formerly a symbol of the samurai order, swords became a standard military accoutrement, albeit one spectacularly unsuited for modern warfare. Officers leading a platoon of soldiers were easy targets when they waved their swords; pilots cluttered their cockpits with swords that served no function; and a few, with time on their hands, used their swords for a variety of cruel or senseless experiments – such as executing prisoners, terrorizing the occupied populations or testing to see if the sword could cut through a machine-gun barrel (it could, but destroyed the blade). The ideals of Yamamoto Tsunetomo proved more enduring that the pragmatism of Tsuchimochi Nobuhide. As a consequence, tens of thousands died rather than recognize the need to survive and fight for another day.

The actions of warriors during the years 1180 through 1615, when many experienced warfare, belie the notion that they were obsessed with death. This account will focus on how warriors fought, and not how their past was idealized. Hence, it will differ considerably from many books about the samurai, which tend to implicitly support later idealizations of the warrior, or favour weapons that attracted interest in the form of schools of marital arts.

Bows and arrows constituted the basic weapons for the samurai throughout the sixteenth century, so much so that warriors referred to themselves as practitioners of the 'way of the bow and arrow.'

LEFT: *In 1877, Saigō Takamori, a leader of the Meiji Restoration of 1868, launched a major rebellion as a reaction to how centralized and authoritarian the new Meiji state was becoming. Saigō and his men relied on guns until they ran out of bullets, when they resorted to using swords. Badly defeated, only a few survived to surrender to the victorious Meiji armies. After the 1877 defeat, this rebellion was portrayed as being a completely anachronistic 'last gasp' of the samurai, when in fact the truth is more complex, for some of the rebels had been inspired by the writings of the French enlightenment philosopher Jean-Jaques Rousseau.*

BELOW: *Japanese soldiers celebrate the fall of the Philippines in April 1942. Notice how many* katana *swords are brandished. These swords, which had fallen from favour in the late nineteenth and early twentieth century, were used commonly in the 1930s and 1940s, but more for display than actual combat.*

The most skilled practitioners could shoot from horseback, while others fought on foot as skirmishers. Only rarely did troops engage in close-up, hand-to-hand combat.

Organized formations of troops rarely existed prior to 1467, instead, foot soldiers scattered before small bands of horsemen, who controlled the battlefield, and who could force any lone swordsman, regardless of their skill, to flee. To classify the fighting forces as armies is a misnomer, for most consisted of dozens, or perhaps, hundreds of men. Command and control rarely existed, as bands of horsemen would advance as they saw fit. Being seen, and thought to be 'worthy of reward' superseded specific

FLAG BEARER

Banners were popular means of identification. They consisted of a white flag with some symbol of a family. The oldest visual representation can be found in Takezaki Suenaga's *Scrolls of the Mongol Invasions of Japan,* dating from the late thirteenth century. Some flags depict Chinese characters, but most show either abstract symbols or stylized images of plants or animals. The family became a distinct entity in thirteenth century Japan, and the oldest such crests date from this time. As warfare became endemic over the course of the fourteenth century, some families fragmented, with brothers adopting new names and crests. Some of these new crests revealed political allegiances. Followers of the Ashikaga, for example, sometimes incorporated the Ashikaga crest (a circle intersected by two black lines) with their familial mark. As armies increased in size in the sixteenth century, banners went from being a marker of each warrior's family to depicting the seal of the lord, or *daimyō*. The illustration here is that of a follower of Katō Kiyomasa in Korea, 1593.

orders of early commanders. Takezaki Suenaga (*c*.1245–1324), who fought against the Mongols for example, completely ignored the commands of his superiors to wait for reinforcements, and instead charged recklessly and was shot off his horse. Nevertheless, he expected rewards for being the first and, ultimately, received them.

MARTIAL FAMILIES

Each warrior house constituted the basic unit of military organization. A familial crest, attached to the armour, identified each unit, although at times this was supplemented by further identification on cloth or silk. A single bannerman, riding on horseback with his unit's flag attached to his back, allowed each unit to be identified from afar. Serving such a role required great bravery, for flags constituted easy targets, and bannermen suffered disproportionate casualties. Many were shot with arrows, or when approaching castles, crushed with rocks. Capture of a unit's flag meant ignominy for the group, and when this happened, these flags were used to taunt the surviving warriors. Unable to put up with such derision, units would, on occasion, engage in reckless counterattacks to recapture them.

During the wars of the fourteenth century, families often divided their allegiances. Some did so knowingly so as to protect their house. Thus one member of the Migita family who was a *gokenin* follower of the Kamakura *bakufu* feigned illness, dispatching his son to fight for Go-Daigo (1288–1339), an emperor who was attacking this regime in 1333. He, to the contrary, fought for Kamakura. Kamakura's loss and ensuing destruction did not ill effect the Migita, for the father who was tainted with his ties to Kamakura merely retired and passed headship of the family to his son, who had fought for Go-Daigo. In other instances, younger brothers with miniscule lands strove for autonomy, and

RIGHT: *Japanese battle flags came in three broad types, as illustrated here. When large enough to be considered standards, they are called* uma-jirushi, *which means 'horse signs'.*

Minamoto no Yoritomo founded Japan's first 'warrior government', the Kamakura bakufu, *in 1185. It remained a limited government, concerned with adjudicating cases among its supporters, and helping to police the provinces so as to uphold the court. Yoritomo, a masterful leader and strategist, was a less successful tactician. He fought in two battles – the first against the Taira in 1180 at Ishibashi was a resounding defeat while the second, in 1189, was against a completely outclassed opponent. Woodblock prints, such as this 'Mirror of Famous Generals of Japan', emphasized his more peaceful pastimes – such as his release of cranes at Kamakura so as to accrue Buddhist merit.*

they did so by offering their support to a commander, and ignoring the authority of their family chieftain (*sōryō*). These people of little means fought most extensively during the civil wars, for it gave them the opportunity to assert autonomy from relatives. Family chieftains proved, however, to be more cautious, in that they desired, above all, the preserve their authority.

Once some rival family members decided to fight for opposing forces, grave problems arose, for familial armour possessed the same crests. Families who fought against each other could suffer disastrous consequences, as allies could not be distinguished from enemies. Some, such as the Yūki, resorted to cutting off a shoulder panel (*sode*) of their armour and attaching it to the 'horns' (*kuwagata*) of the helmet, while another family in similar circumstances decorated their armour with small bamboo leaves. Unsurprisingly, many new familial crests were created over the course of the thirteenth and fourteenth centuries as some warriors asserted their autonomy as a 'new' family.

WEAPONS OF WAR

During this early, scattered style of fighting, warriors relied almost entirely upon bows. Swords would constitute weapons of personal defence, to be hung above one's bed, in case of robbery, so as to be easy to repel attackers, and could be used to stab opponents who ventured near, but they were not commonly used in battle. This changed, however, once civil wars engulfed the archipelago from 1333 through 1392. With the onset of civil war in the early fourteenth century, foot soldiers began to hold their own against mounted archers. They did so by occupying swampy or mountainous terrain, where horses could not roam. Some favoured long swords to slash the legs of nearby horses, and these weapons proved effective when confronting single horsemen, but were less helpful when confronting a squad of archers, who could surround and shoot to death even the greatest swordsman. Foot soldiers would not dominate the battlefield until they could be trained and organized in tight formations, which only became possible in the fifteenth century.

Once men could be trained in close formations, the pike became a favoured weapon, for it was cheap and could be used to great effect by a massed force, or phalanx. These formations, composed of lightly armoured warriors, could occupy contested lands, and with their wall of pikes and shields they could defeat horsemen on level terrain. This tactical innovation transformed fighting styles in 1467, as men on foot rather than mounted archers dominated the battlefield.

INNOVATIONS

The Ōnin War lasted for ten years (1467–77), and resulted in heavy casualties and the devastation of the capital, Kyoto. From a military perspective, the conflict is as much one of innovation in tactics, and logistical prowess, than unbridled destruction. For ten years, men and supplies were pumped into the battlefield, as armies fought for control over a few miles of land. Much of the capital was burned in order to give horsemen space to roam, but the advent of the phalanx meant that even on burned-out ground, horsemen could no longer stand against a force of pikemen. The situation resembled that of the Western Front during World War I in that massive trenches were built, and watchtowers looked over enemy fortifications. Pikemen and skirmishers fought to occupy strategic positions and suffered casualties while doing so, while cavalry became limited to conducting peripheral raids, rather than constituting the primary force on the battlefield.

With the ascendance of defensive tactics, and the annihilation of the capital, generals strove to create as large

ABOVE: A basic archery pose. Notice how the bow is held not in the centre but the lower half, and likewise note the grip and way of holding the arrow shaft between the index and middle finger.

RIGHT: A warrior procession surrounding an emperor's cart, from the Heiji monogatari emaki *(Heiji scrolls). A noble, Fujiwara Nobuyori, depicted to the left in court robes, leads this procession. As late as 1159, the time of the Heiji insurrection, samurai remained subservient to the court.*

an army as possible. In the century after the Ōnin War armies increased by a factor of ten or, if we are to trust the numbers, a hundred, as regional magnates effectively mobilized most of the male population of their areas. With the expansion in the size of the army came uniforms, flags and above all pikes. As time passed, these weapons nearly quadrupled in length.

Nevertheless, arrows remained the dominant projectile until late in the sixteenth century. Guns arose in Japan in two waves, the first occurring in 1466, on the eve of the Ōnin War, as primitive three-barrelled weapons that impressed with their sound, but not their penetrating potential. The Portuguese introduced harquebuses in 1543, which proved more effective. Priests from the temple of Negoroji quickly gained hold of these weapons, and soon constituted a

LOGISTICS

Logistics remained the great obstacle to organizing larger forces. Warriors commonly complained of the expenses of staying in the field for extended periods of time, for weapons had to be repaired and replaced, along with horses, and the demand for victuals remained constant. Often, many would depart from their encampment and return home for they could not pay for indeterminate service.

Those who were skilled at raising money, however, gloried in their wealth, and flaunted it by dyeing their horses in bright colours, wearing mighty crimson capes and placing gold and silver fittings on their swords. Commanders were granted the right to use half of a province's revenue for supplies in 1352, and this allowed them to gain great power and rapidly outpace their fellow warriors. These men, known as 'protectors' of a province, or *shugo*, used the revenue of the state to build castles, establish foundries and conscript warriors. As they gained control over lands, they gradually forced housemen to become their retainers, or samurai, although this process would take centuries. And as they gained wealth, and power, they could start to train standing armies.

*Hirosaki castle, originally built in Hirosaki to the far north of Japan in 1611. The central five-storey tower (*tenshukaku*) burned down as a result of a lighting strike and was replaced with a much smaller three-storey structure in 1810. This building was designed to withstand an attack, even though Japan had witnessed no warfare for centuries. Note how the structure has no windows, with instead slits for archers and gunners. Boulders could be stored in the lower section, which overhangs the stone walls, and dropped on attackers.*

A nineteenth century woodblock print by Yoshitoshi, entitled Biographies of Valiant Drunken Tigers, *refers to valiant swordsmen who fought in the battles that accompanied the fall of the Tokugawa. Some, influenced by schools of martial arts, had come to believe that the sword represented the most effective combat weapon, but these figures were decimated. This image shows a doomed fighter who could effectively chop a food tray in two, sending objects flying, as he rushed to battle.*

formidable force of gunners, while the Ashikaga shoguns also transmitted knowledge of these weapons and gunpowder to supportive warlords. Gradually, these guns supplanted bows. They inflicted heavy casualties during the battle of Nagashino in 1575, although guns and bows continued to inflict wounds at comparable rates through the 1580s. Not until 1600 did guns cause a majority (80 per cent) of projectile wounds.

After the major battles of 1600 and 1615, Japan settled down to a period of fitful peace and stability, dominated by the Tokugawa family and marked by, as we have seen, the establishment of a samurai order and urban centres.

Samurai received stipends in rice and not cash, and dramatic increase in production allowed for prosperity, but caused warriors to suffer diminishing purchasing power and relative impoverishment. The cult of frugality and privation, very different from the early ideals of wealth and ostentation, served the samurai. Some became so poor that they had to pawn their swords and armour and engage in small-scale manufacturing.

With the establishment of peace, the Tokugawa regime did not abandon the gun; rather they monopolized the production and dissemination. The production of cannons was similarly carefully controlled. Japan severely limited its contact with European countries, but remained aware of developments.

Some individuals studied Dutch learning, as it came to be called, and one samurai, Sakuma Shōzan (1811–64), experimented with European cannon, guns and new military furnaces in 1841, while in the domain of Saga, a blast (reverbatory) furnace capable of smelting iron, which required a temperature of 1300°C (2372°F), was produced in 1850, and manufactured working models of the telegraph and steamship in 1852.

THE END OF THE SAMURAI

Some samurai embraced cultural and technological change, while others, such as Eto Shinpei who were reformers in other ways, actively resisted it. Nevertheless, as Japan modernized, there was no place for an idle 6 per cent of the population (an 1850 survey put them at 1.8 out of 30 million) to be paid by government stipends. These were converted to bonds, allowing the wealthiest samurai to become capitalists in the new regime, while their poorer brethren adopted a variety of professions: teacher, newspaper editor, or, looking for a good use for their swords, barber.

After Saigō Takamori's last major rebellion, which was as much a struggle for popular representation as it was for a return to the past, the samurai ceased to exist, but their ideals, if not fighting practices, would be ensconced in the Japanese national psyche in its educational doctrine through 1945. Nearly all of the Japanese came to glorify values of the samurai until the cataclysmic debacle of the war caused this last residue of the earlier samurai ideals to wither as well. In the Japan of today, the actual weapons and fighting techniques of the samurai are no longer well known or widely understood.

Swords symbolized the samurai warrior, while pikes remained weapons that had no connotations of any link with status. Notice in this kabuki *play, how a* katana-*wielding actor defeats another holding a pike. In actual battles of the sixteenth century, the pike was the favoured weapon.*

Mounted Samurai

By the year 1200, a distinct military order had arisen consisting of most, if not all, of the mounted warriors of Japan. For several centuries prior to 1200, these elite members of provincial society had served as deputies to provincial governors, and through these ties they gradually began to assert the ability to manage the land, be it the public domains of the province, or in estates that had been commended to members of the nobility. Their position remained precarious, as their right to manage estates depended on the whim of those in authority.

Mounted samurai armed with yari *(spears) and distinctive* sashimono *(streamers) for identification charge in an action shot from Akira Kurosawa's movie* Ran *(1985). The armour and weapons reflect late sixteenth styles.*

This situation bred resentment, which in turn caused many of the warriors to revolt against the court and their delegated representatives, the provincial governors. Minamoto Yoritomo (1147–99), a commander of impeccable warrior lineage who had been an exile since the age of 13, launched what can perhaps be characterized as a conservative revolution, for he rebelled against central authority in 1180 but he did so by strengthening warrior rights to the land. In particular, he granted the office of land steward, or *jitō*, to his supporters, and this office served profoundly to influence their identity and behaviour for the next half a millennium.

In 1183, Yoritomo went from being a rebel to an upholder of the laws of the court. Having secured the right that he had unilaterally asserted since 1180 to appoint his followers to the post of *jitō*, he then used his authority over provincial warriors to force them to abide by the edicts of the court, and to abandon other depredations or outrages. Having earned legitimacy, he now nullified the *jitō* posts for all those who claimed to be *jitō* but did so without his investiture. From this time until his death in 1199, Yoritomo then spent much of his time quelling uprisings and confiscating these posts

from recalcitrant figures, or those who refused to serve him in a campaign of 1189.

LITIGATION AND LAND

The need to adjudicate disputes and determine, or legitimate, inheritance patterns, proved the defining feature of Japan's first warrior government, the Kamakura *bakufu*. Kamakura was above all a judicial government, concerned overall with lawsuits, as the samurai of the thirteenth century were among the most litigious people on earth. Disputes, some of which would drag on for decades, were generally judged fairly, and warriors had enough confidence in the system that they only infrequently resorted to bloodshed. Law represents one of the greatest legacies of this warrior government, and its code, the Jôei formulary which was penned in 1232, provided far greater protections of land rights for a greater cross section of society than the Magna Carta of 1215.

In addition to adjudicating disputes, Kamakura was responsible for providing order in the capital and provinces. Through regional constables, or *shugo*, appointed to each province, Kamakura requested that its warriors, known as *gokenin*, or 'honourable housemen',

ABOVE: *The photo is a famous portrait, until recently thought to represent Minamoto no Yoritomo. In it he wears court robes and has the fittings of a captain of the imperial guard for his sword. Recent scholarship suggests, however, that this portrait was actually Ashikaga Tadayoshi, the younger brother of the first Ashikaga shogun.*

LEFT: *This scroll, created during the thirteenth century, depicts a coup attempt by Minamoto Yoshitomo and Fujiwara Nobuyori in 1159. Note the box-like* ōyoroi *armour of mounted warriors, as well as the simplified* haramaki *type armour worn by men on foot. Older armour provided little protection for the lower body, particularly for low-ranking warriors. A face guard called a* hatsumuri *protected the foreheads and cheeks of most warriors. Note, too, how mounted warriors almost exclusively relied on archery while foot soldiers used swords and glaives (*naginata*) to decapitate an enemy prisoner.*

An official of the Kamakura regime wearing robes emblazoned with the familial crest of the Hōjō. This image, from the Scrolls of the Mongol Invasions, *which was created late in the thirteenth or early in the fourteenth century, is the oldest depiction of such a crest, and reveals how familial identities were becoming more pronounced in the latter half of the thirteenth century.*

provide guard duty or, in times of need, military service. Samurai men and women, such as the nun Hō Amidabutsu, were eligible to serve, and this duty was not confined to those who held the post of *jitō*. Guard duty proved burdensome, for it entailed long periods of service, and *gokenin* were liable to pay for other levies, such as labour to repair broken dykes. Some warriors preferred to abandon the designation of *gokenin* because of the burden that this status imposed, although this changed in the late thirteenth century, when all *gokenin* were granted debt relief by the Kamakura *bakufu*.

BENEFICIAL POSITION

The office of *jitō* proved beneficial for a warrior appointed to this position. Taxes still might flow to proprietors, and cultivators continued to work the land, but the *jitō* served as manager of the land, who could police the areas and impose ad hoc levies, thereby enriching himself in the process. As long as a *jitō* was not implicated in rebellion, he could maintain these lands and pass them to whomever he saw fit. Ability, rather than primogeniture, served as the general principle of inheritance, and although some lands were passed to a single heir others were conveyed to multiple sons and daughters. One warrior, who otherwise had no heirs, even passed his lands to his two faithful hunting dogs, named 'Big Black' and 'Little Black'.

Ownership of specific parcels of land, or, to be more precise, the ability to manage the cultivators of specific plots of lands, profoundly influenced the samurai. Many adopted the name of the lands where they lived, thereby revealing how closely they identified with their holdings. At times, several brothers would change their surnames, each taking the name of his landholdings. They did so in order to assert that they were autonomous, and not beholden to their brothers, or uncles. Concurrent with this change, many new families appeared, each with their own 'house' and familial crest, which began decorating warrior clothing for the first time during the mid thirteenth century.

The ability to manage the land, and receive an investiture to the office of *jitō* proved more significant than the ability to ride a horse, which many could do. High levels of literacy existed among samurai men and women, because they had to write wills to pass their lands to their heirs. Nevertheless, the ability to ride a horse constituted an important social marker. Those who could ride constituted the elites of provincial society. Possessing enough resources to raise and maintain horses they could, if need be, impose their will on cultivators, for few on foot could stand up to even a small squad of horsemen, particularly those capable of shooting arrows while riding fast.

LORDS AND FOLLOWERS

The horse remained fundamental to military organization through 1467, so much so that military units were invariably counted in terms of horsemen, while those on foot did not merit mention until midway through the fourteenth century. The divide between those who could ride, and those who could not, provide significant. In 1284, for example, Japan's warrior government, the Kamakura *bakufu*, for example, prohibited priests and 'the base' from riding horses. Even though priests and commoners could not ride with sanction, others who lacked the recognition of the post of *jitō* still thought themselves the equals of warriors, leading to considerable tension in Japan throughout the year of 1333.

Of those who could ride, the most illustrious were warriors with lineage and lands. They rode the mightiest horses, and were readily identifiable with lavish box shaped armour. These *gokenin* possessed an expansive residence and lands, and would be responsible for providing for a group of followers, squires as it were, who ranged in number

from three to 35. Their followers were called *samurai* in contemporary parlance. These men had their horses and armour provided for by their *gokenin* lord, and if they were wounded or killed in battle, it would be their lord who was compensated, not themselves. Regardless of what great deeds any samurai accomplished, they would not receive individual recognition. Instead all honour and rewards accrued to their lord.

Some warriors lacked any specific legitimacy, or sanction from the Kamakura regime. Located more in central and western Japan, they did not possess *jitō* rights, nor were they called up on guard duty. These men, called *myōshu*, or 'lords of the *myō*' (the *myō* being a unit of land), considered themselves the equals of *gokenin*, hence they were loath to serve under them. Briefly in 1281, manpower shortages arising from the need to defend against the Mongols caused some Kamakura *bakufu* officials to attempt to incorporate *myōshu* into their forces, but this policy was quickly abandoned.

With the outbreak of civil war in the 1330s, however, many were called up to fight, including both *gokenin* and *myōshu*.

A bannerman of Takezaki Suenaga rides a horse in front of a stone wall built to protect Japan from the Mongol invaders of 1281 and carries Suenaga's flag. In contrast to the warriors who sit on top of the stone wall, he does not have shoulder armour (sode) to protect his upper arms. He is wearing simplified haramaki armour.

The collapse of Kamakura in 1333 and the continuation of war for two generations allowed these two distinct orders to coalesce into a single, larger social unit, whereby samurai were known as 'men of the province' (*kokujin*) or 'outsiders' (*tozama*), the latter term emphasizing their autonomy. In some regions, a small band of horsemen, too poor to otherwise achieve autonomy, banded together as a collective unit. They were known as *hyakushō*, or the hundred names, a term that mistakenly has been thought to denote peasants. As they had a name, they occupied the fringes of elite society, for they were substantially better off than the next, and most common strata of society, which was identified as 'the base' (*genin*). Remarkably, *hyakushō* rode horses and fought in battle, and received recognition as a single band. Over the course of the fifteenth century, and the rise of great armies, this autonomy became impossible, and many joined warrior organizations as dependent followers (*samurai*), while others focused more on cultivation. These latter individuals were readily distinguishable, however, for even though they rode horses they had simplified armour, often nothing more than a stomach guard (*hara-ate*).

Having provided an overview of the nature of warrior society, let us now explore how the samurai of the thirteenth and fourteenth centuries fought. And in understanding the fighting techniques of the samurai, nothing proved more fundamental than the enduring partnership between man and horse.

VALUED CREATURES

The samurai treasured their horses. So important were horses that specially designated areas of pastureland were established throughout Japan, with the best horses coming from the pastures of eastern and northern Japan. Horses were valuable, for they cost about half the price of a suit of armour. In the fourteenth century, they can be documented as being worth three of four kan (3000–4000 dollars or 1500–2000 British pounds). They remained an asset, however, and some warriors can be documented as selling their horses in times of need. Other proud owners dyed their mounts in hues of crimson, purple, chartreuse and sky blue, or added stripes, so as to highlight their steeds' magnificence. Other commanders preferred showing off their mounts by relying on tigerskin saddle blankets, which involved considerable expense, for they had to be imported from the Asian mainland.

A depiction of a Japanese horse with a tigerskin saddle blanket. This illustration suggests that the Japanese horses were more majestic than in fact they were, for they were the size of ponies.

THE CHARACTER OF JAPANESE HORSES

Intelligent, independent and stubborn, the Japanese horses prized their autonomy, much like the warriors who rode them. A rider used to more docile mounts would describe them as ill-tempered brutes, for the Japanese horse would not unquestioningly abide by the commands of its master, but most samurai seem to have tolerated this independence.

No tradition existed of gelding horses, which is unusual, for most people who rely heavily on the horse do so, and this contributed to the disorganization of early battles. Stallions fought when housed in close quarters, and the presence of mares in heat only served to exacerbate tensions among the horses. An enemy encampment with mares in heat could have disastrous consequences for an unlucky samurai riding a frisky stallion, for he would have to either dismount or ride alone amidst the enemy with an otherwise preoccupied horse.

The Japanese horse is a subspecies of the Mongolian horse, although some specialists believe that they resemble most closely primitive horses, now extinct, such as the Tarpan. Horse is perhaps a misnomer, because by modern classification all but the largest of these

LEFT: *The tarpan, a primitive and now extinct horse that resembled Japanese horses of the samurai era.*

BELOW: *A* yabusame *archery competition, whereby riders on a galloping horse fire at several wooden targets to their side. Such competitions have been waged for centuries. The horses now used, and photographed here, tower over native Japanese horses, for it was the policy of the Japanese army in the nineteenth century to improve the size and strength of Japanese horses by breeding them with European cavalry horses.*

beasts stood at 140cm (slightly over 4ft 6in or 14 hands), and would still be classified as a pony. Minamoto Yoritomo rode a mighty beast, standing at 142cm (4ft 10in or 14.2 hands), which currently just barely passes the dividing line between horse and pony, but his mount was exceptional. An excavation of horse burial grounds dating from the fourteenth century, reveals that most horses stood only slightly over 130cm (4ft or 13 hands) at the shoulder with the smallest horses being only 109cm (3ft 6in or under 11 hands), which is standard donkey size. Most Arabians, by contrast, stand approximately at 152.4cm (5ft or 15 hands) at the withers, while thoroughbreds average 162.56cm (5ft 4in or 16 hands).

The short stature of Japanese horses also explains why there is no tradition of heavily arming these beasts with plate armour, as arose in Europe. On occasion, some kind of chain mail seems to have been used in the fourteenth century, while some *daimyō*, such as the later Hōjō, stipulated that their mounted warriors should cover their horses in armour in the sixteenth century. None attempted to joust with lances, or to use their horses to shock and smash their opponents, because their mounts would not, and indeed, could not, do so. Horses

Tomoe Gozen, a famed and presumably mythical woman warrior who was mentioned in the Tale of the Heike, *an epic that was completed in the fourteenth century.*

hooves were not protected with horseshoes as well, for they would not be adopted until the mid eighteenth century, when increased knowledge of European practices was disseminated via Dutch texts. Instead, they were equipped with straw sandals, which provided minimal protection for their hooves, and closely resembled those worn by their samurai riders.

Diminutive horses placed large men at a disadvantage on the battlefield. Fujiwara Kunihira was a very large man, and his size prevented him from skilfully riding what was known as the 'fastest horse of northern Japan'. He was a scion of the northern Fujiwara, who ruled the provinces of Dewa and Mutsu from their city of Hiraizumi for four generations. We do not know Kunihira's height, but those of his mummified relatives exceeded 1.8m (6ft). His poor mount, which, standing at 141cm (4ft 9in or 14.1 hands) was actually a small horse (and thus very large for Japan) 'broke out in a sweat' every time he rode it up the highest hills of Hiraizumi. Kunihira was killed rather ingloriously, unable to ride his horse well in battle, on the tenth day of the eighth month of 1189.

Women warriors, by contrast, had an advantage as riders, as they were generally lighter and more agile than men. Those women of a warrior household would have learned to ride, and one can find references to them fighting on horseback alongside men. The *Tale of the Heike* recounts the exploits of one named Tomoe Gozen, who has become the most famous woman warrior of Japan. She is suspected of being a fiction, but the fact that women fought can be documented in more reliable sources. In 1351, a force of predominately female cavalry can be documented as fighting in western Japan, and some suits of armour tailored to the female anatomy survive to this day. Warring women were not particularly common on the battlefield, but at the same time, their participation in combat was not rare enough to generate much surprise.

SPEED IN COMBAT

Being short, and stubby of leg, Japanese horses were incapable of great speed. An experiment by Japan's NHK public television network in 1980 revealed that a war pony carrying a rider in full armour could never surpass a speed of 9km/h (5.6mph). For this test, a pony of 130cm (4ft 3in) was selected, weighing 350kg (770lb). A total of 95kg (209lb) of weight was added, with 45kg (99lb) accounting for the armour and saddle, and 50kg (110lb) accounting for the weight of the rider. This poor mount, at

Members of a Mongol historical re-enactment of Genghis Khan's armies galloping on the steppe. Their horses closely correspond in size to those ridden by the samurai, and are surprisingly small. In spite of their size, they gallop smoothly. This trait helps explain the accuracy of Mongol archers.

its fastest, cantered (*kake-ashi*) but could not sustain such a pace long before dropping into a trot (*haya-ashi*). Tellingly, the word canter most commonly describes horses in battle, with references to a gallop (*shikku*) appearing very rarely. Although modern ponies in Japan may not be of the same physical conditioning of their ancestors, the slowness of a horse must also be taken into consideration in reconstructing battle. Galloping was reserved for short sprints, or desperate situations, but otherwise a cluster of samurai would cross the battlefield on trotting, or cantering mounts. Such slowness would seem to take away the drama of the battlefield, but conversely it would allow for more accurate archery.

These slow horses did have their advantages. The excelled at surmounting broken terrain, something that is important in Japan, which is 80 per cent mountainous. In one famous episode from 1184, Minamoto Yoshitsune (1159–89) led a small force down the side of a steep hill at the battle of Ichi-no-tani, thereby surprising his enemies and defeating them. Longer legged horses could not master such a descent.

The second advantage that they had is that, like their brethren, the short Mongol horses, they run extremely smoothly and do not jolt the rider, thereby allowing for accurate shots to be fired. Although cantering proved less speedy than the gallop, it proved far more sustainable, and was likewise far more suited to archery than the much bumpier trot. This slower pace allowed horses generally to avoid being mired in poor terrain, such as muddy paddies, but they were not infallible, and during winter campaigns, some horses riding

A warrior from the Illustrated Scrolls of the Events of the Heiji Era. *This warrior wears impressive armour made of black lacquered plates bound with red, black and yellow chord, highlighted with a gold sword, and gilt decorations on his helmet and shin guards. In spite of the munificence of his dress, this man did not wear the headgear of a general and should not be confused with one.*

over snow could break through hidden ice underneath, and be trapped in swamps, paddies or rivers.

SADDLES AND TACK

Analysis of horse tack suggests that samurai most prized stability in the saddle, rather than speed. The horse, and by extension, the saddle, were designed to provide an immobile platform for archers to shoot their opponents. Saddles also protected the lower torso of the rider, but this heavy, large box-like structure that rests on the back of the horse was not designed with the comfort of the horse in mind. Most saddles were made of lacquered wood, which meant that the wood was treated with resin drawn from a plant possessing the same unfortunate qualities as poison ivy. This resin hardened the wood, and made it impervious to rot, which is why saddles and, for that matter, many armour plates were constructed from lacquered wood. Lacquer had the advantage of looking attractive, for it dried in a sleek smooth black finish that could be embellished with gold or silver decorations.

These saddles were complex structures, which required considerable time to be placed on a horse's back. First, an under-saddle (*shitagura*), which functioned as a saddle blanket, was first placed on its back. This under-saddle could be made of padded leather or furs as exotic as tiger skins, which had to be imported from Korea or China.

On this under-saddle, a wooden frame, the saddle (*kurabone*), would be attached by being tied tightly with hemp rope. The wooden saddle consisting of two wooden bands (*igi*), which would sit parallel on either side of the horse's backbone, and two interlocking boards that would attach to the front and rear of the *igi*. These boards, called *maewa*, functioned as a pommel board, and a *shizuwa*, or cantle, to the rear, completed the saddle.

The pommel board and cantle block represent the defining feature of the military saddle (*gunjingura*), for they were unusually deep and heavy, so as to

Fujiwara Nobuzane provided a highly accurate depiction of the several guardsmen of the retired monarch Go-Saga in his Mounted Imperial Bodyguards scroll (Zuishin teiki emaki, *c. 1247). He took great care in drawing horse accouterments such as the elaborate girth, chest strap* (munegai) *and rear strap* (shirigai) *as well as the stirrups, bit and halter.*

SADDLES

Saddles were designed to be stable platforms on horses. They allowed for little movement which aided accuracy in archery but did not contribute to the speed of the admittedly slow Japanese ponies. The *igi* boards were placed above the saddle blanket, and in the front one sees a *maewa* which functioned as a pommel board, while the backboard is called a *shizuwa*.

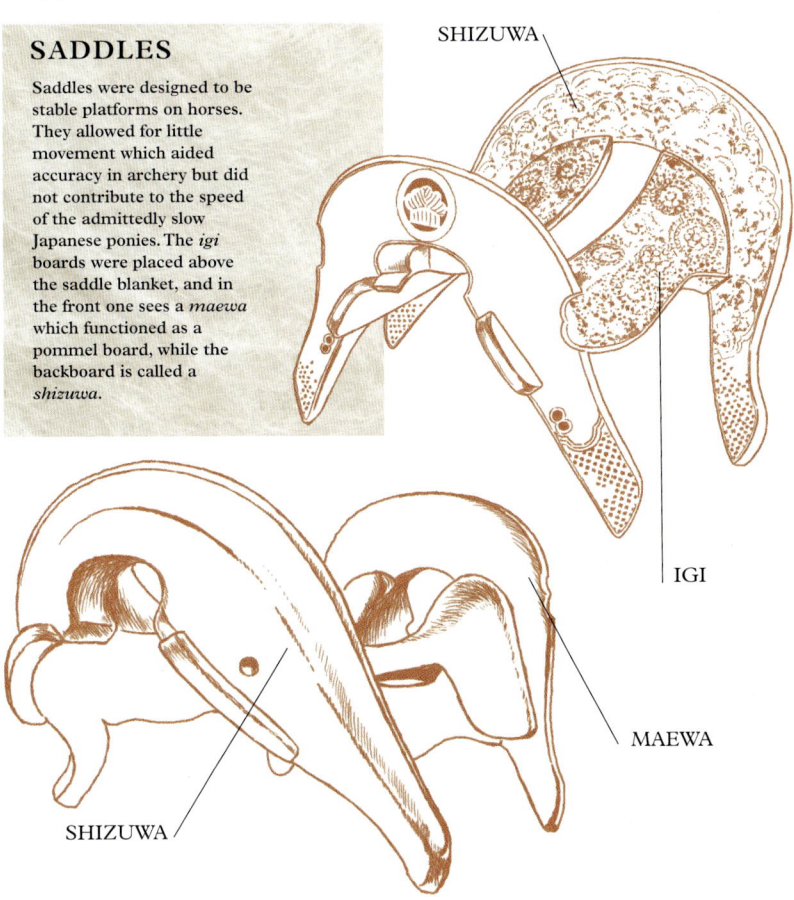

STIRRUPS

Another distinctive feature of horse armaments, the stirrup, has a long heritage in Japan, with simple metal rings being used as early as the fourth century. The military stirrup was like a cup, made from lacquered wood, and with the bottom elongated so as to allow most if not all of the foot to rest firmly on it. This stirrup allowed for ease of standing, and had the advantage of making it nearly impossible for a dismounting rider to accidentally catch his foot and be dragged by a frantic horse. The 'dove's chest' (*hato mune*) prevented the toes and front of the foot from being injured. In all, the thickness of the wooden stirrup, coupled with the depth of the saddle, served to protect the mounted rider, whose upper body was covered by a unique and effective suit of armour.

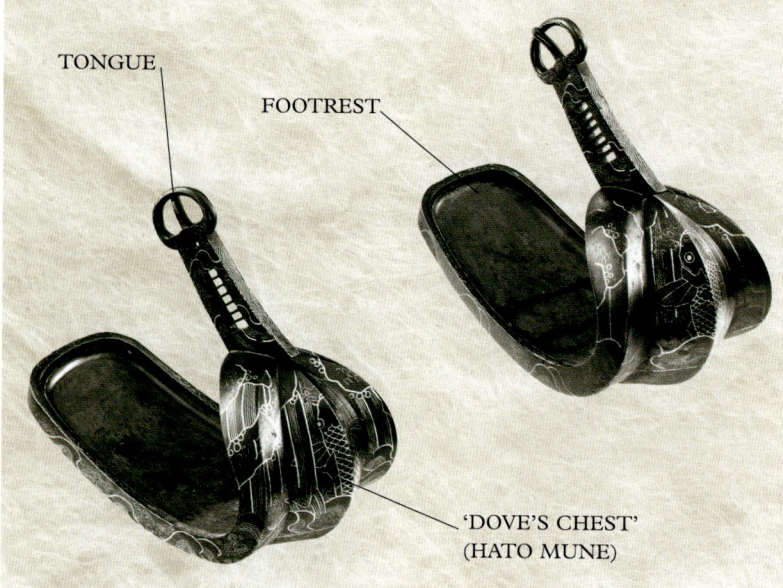

TONGUE
FOOTREST
'DOVE'S CHEST' (HATO MUNE)

protect the lower torso of the rider. The pommel boards were set astride a horse's withers, the high spot between its shoulders, while the cantle block rested on the rise of a horse's lower back, and served to protect a warrior's rear. These two wooden boards were carved to fit into the *igi*, and tied together as well, so as to make the structure as rigid as possible. This structure would be tightly bound so as to ensure that it would not slip at all.

In addition to tying the under-saddle to the wooden frame, a cinch or girth would be wrapped around the horse's belly, threaded through the under-saddle, and through slots in the wooden bands, where it would be tied. Above this girth knot, on top of the wooden structure, a padded seat (*basen*) would be held in place by stirrups, which were connected by leather and threaded through slots in the igi and holes in the under-saddle. Silk or folded fabric would then tie the pommel board to the chest of the horse. This chest strap (*munegai*) would be matched by another rear strap, the *shirigai*, which attached the cantle by wrapping around the horse's hindquarters (the croup) and around the dock, the fleshy base of a horses' tail.

All straps, including the reins, would be made of hemp, folded fabric or silk; leather, common in Europe, was rarely used in Japan. Two set of reigns were used, one connected to a halter for leading the horse while dismounted, and another set for steering the horse, which was attached to a bit. The horse was controlled by the bit in its mouth, which was made of steel and attached to two cheek pieces, and steel rings attached reins to these cheek pieces. Upon mounting his steed, the rider invariably tied the reins used for leading his horse, which were attached to the halter, to the pommel board. The second set of reins were used to steer and stop the horse, but in times of battle, when a rider needed to shoot from horseback, they were either tied or loosely attached to the pommel board so as to allow the archer to aim and shoot while riding. This action is less precarious than it might seem, for as we have seen horses advanced at a more leisurely pace than has been imagined.

While shooting, the archer most commonly shot to the side or to the rear of his horse. Care needed to be taken not to wave objects near the horse's head. Inexperienced or flustered riders who drew a sword, for example, could be dropped from their horse. Picture scrolls such as fourteenth-century *Kasuga gongen kenki e* depict warriors on horseback with longer weapons, such as a curved blade on a pole, known as a *naginata*, or hooks known as 'bear claws' (*kumade*), revealing that it was possible for some skilled riders to wield hand-held weapons on occasion.

THE ARMOUR OF THE MOUNTED SAMURAI

Most early Japanese armour was specifically geared to protect against arrows, which constituted the dominant weapon of thirteenth- and fourteenth-century warfare. The distinctive armour of the samurai was ideally suited to protect a rider on horseback. Bulky and box-like, this loose-fitting armour had hanging sections, a skirt of sorts, which hung over the saddle, and further protected the mounted warrior. While on horseback the rider otherwise required little protection for the legs and feet, because as we have seen, the saddle complemented the armour and protected the lower body of a warrior.

Flexibility and lightness, and thoroughness of protection, represent the two opposing constraints of all armour. In different times, different cultures have preferred and different weapon sets demanded that armour of contrasting

ŌYOROI STYLE ARMOUR

A reproduction of *ōyoroi* armour, presented by the last Tokugawa shogun, Yoshinobu, to Queen Victoria in 1860, that was manufactured by Masuda Muneharu. Muneharu's suit contains many panels of laquered wood, laced together, as is consistent with the earlier suits of armour, and also has a boxy section protecting the chest, with archaic pieces like the *sendan no ita*, *waidate* and the *kyūbi no ita* attached. The helmet has prominent *kuwagata* horns and a pronounced *fukikaeshi* to either side of the face. Nevertheless, Muneharu was unable to construct a historically accurate reproduction. The prominent mask did not exist in earlier times, nor did the neck protection or *nodowa*. The *sode*, which protected the upper arm and shoulder, are flexible here, but in the original were inflexible and acted much like shields. Likewise, the *kusazuri*, or woven legging protection, is much longer and less box like than it should be.

types be used. One could chose protection, which entailed wearing more metal, and thereby sacrificing lightness and flexibility. In medieval Europe, chain mail, for example, proved popular at the times of the Crusades, but even the weight of this flexible armour had its drawbacks, as Frederick Barbarossa (1122–90) discovered in 1190 while leading an army in Anatolia, when he drowned in the Saleph Creek in water that was not waist high. Frederick's ignominious end notwithstanding, the Europeans preferred strong protection, ultimately favouring heavy plate mail to chained links, with a corresponding decrease in mobility.

By the fifteenth century, these disadvantages weighed heavily, as French knights who fell off their horses, as many did in Agincourt in 1415, could not arise from the mud, or readily defend themselves against lightly armed opponents who used broad knives to stab through the joints of their armour and mortally wound these supine knights.

CLIMATE CONSIDERATIONS

The climate of Japan is among the wettest for regions outside of the tropics. An extended rainy season, coupled with typhoons, means that Japan receives on average 170cm (5ft 6in) of rain a year, which is twice the world average, with some areas of Kyushu receiving almost 2.29m (7ft 6in) of rain. In such a humid climate, iron rusts easily, and the earliest armour, used as early as the fourth or fifth century, survives only as hulks of rust in tombs, or protected with resin and now restored in museums. In Japan, the armour of the samurai is commonly known as 'great armour' (*ōyoroi*). It is box shaped, with prominent sleeves, or *sode*. The armour was made of thousands of tiny plates, called *sane*, which were made of leather, and then coated with lacquer. These strips were then woven together with bright cloth. Often, in the chest region the lacquered wooden *sane* were reinforced with metal plates, so as to prevent a lethal shot to the chest.

The cloth used to weave armour plates together could be dyed in a variety of colours, most commonly red and white, and the pattern itself could be used to identify a particular warrior family. Unlike chain links, or metal plate armour, damaged sections could be rewoven relatively easily on the battlefield, and, if need be, certain *sane* added. Rain did not damage this armour, as it was mostly constructed of leather and lacquer, although its cords could rot or house fleas and lice.

The best armour was created in the capital, as the Kamakura *bakufu* official Kanezawa Sadaaki noted in a letter.

DŌMARU ARMOUR

Dōmaru, constituted a simplified style of armour that was attached in the back. Like the older armour, small lacquered plates were woven together, but the armour was less box like than earlier examples. This version is typical of suits dating from the fourteenth century. It includes *sode* but many did not. Note too the *nodowa*, or throat protection, and the fact that the helmet contains decoration, in this case a *maidate*, which became common and was no longer confined to generals.

MAIDATE

To get a relative sense of its worth, it cost four times as much as a simplified suit of armour, and eight times as much as a sword. It required two years to construct one suit of armour, which involved the weaving of 2000–3000 *sane*, although in times of emergency some suits could be constructed in as little as a year. In times of high demand, particularly in the fourteenth century, some suits of armour were largely constructed from leather.

The *sode* constituted one prominent feature of this armour. These sleeves functioned as portable shields, for they protected the sides of a warrior as he was shooting arrows. Picture scrolls show how arrows commonly embedded in this box-like structure. Warriors only wore

The scene below from the Mongol scrolls reveals that most, but not all warriors wore eboshi *underneath their helmets, although the man in the centre stated that it was the tradition of his family not to wear them while battles were waged. The man on the left is in informal attire, with only a gauntlet (*kote*) protecting his left hand. All three seated warriors wear fur shoes (*tsuranuki*).*

EBOSHI

Eboshi were an essential headgear for adults, and the act of wearing an eboshi demarcated adulthood. Eboshi were worn by nobles, warriors and commoners. They were made of dyed silk and many were lightly lacquered. Starting in the twelfth century, non-nobles wore softer versions, but after the fifteenth century, very hard standing *eboshi* became popular. Some helmets were later devised which resembled folded or standing *eboshi*.

DRESSING FOR BATTLE

ROBES

Simple *hitatare* robes were worn under trousers with draw strings under the knee. Over fabric socks (*tabi*) straw sandals were worn. They became popular in the fourteenth century because they could be easily replaced and did not house lice and fleas like fur boots. Finally shin guards are attached.

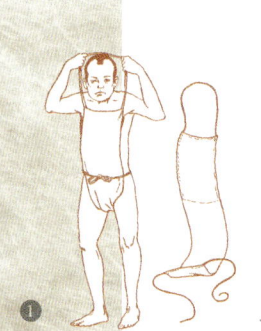

ARMOUR

Kusazuri were attached to protect the legs, and then *kote* gauntlets were added. After adding padding to the upper arms armour was hung over the shoulders and secured with a complex system of cords.

WEAPONS

Finally *sode* sleeves were added, along with swords, a *nodowa* to protect the lower face, and then a cloth band (*hachimaki*) was wrapped around the head, and face mask attached before the helmet was worn. The armour and accoutrements depicted here were used in the fifteenth and sixteenth centuries.

MOUNTED SAMURAI

④

⑤

⑥

⑩

⑪

⑫

⑯

⑰

⑱

43

HEAD PROTECTION

this 'great armour' while on horseback. On foot, the large sleeves proved a hindrance in combat and were little more than useless.

The helmet (*kabuto*) was generally made of iron, with overlapping *sane* protecting the sides and back of the neck. The attached plates could, if we are to believe literary accounts, be ripped off helmets, but they generally provided good protection, particularly for the back of the head and neck. A small visor existed at the front of the helmet. The helmets too were heavy, so much so that prominent samurai often did not wear them until they were going to battle. Some helmet bearers followed their lords, handing over their helmets right before the charge. On other occasions, warriors would exchange helmets with others, so that their actions could better be viewed on the battlefield.

Helmet liners did not come into existence until the fourteenth century. Prior to that, pointed soft hats called *eboshi* were worn, and according to some warrior fashion protruded from a small hole in the top of the helmet. Some variation existed. According to the *Scrolls of the Mongol Invasions of Japan*, warriors of the Kawano family boasted that they would not wear formal *eboshi* headgear while battle was being waged, and some warriors preferred padded caps under their helmets instead. Adequate insulation proved necessary, for a sharp blow to the head could make a samurai unconscious, which made him easy prey to enemy on the battlefield.

Two large horns, the *kuwagata*, were thought to provide the wearer with mystical powers, which is why prominent warriors wore them. They do not seem to have any physical function, except to reveal the more prominent warriors. Generals wore, at times, even more elaborate headgear, with lions protruding from beneath the *kuwagata*, as in the case of a thirteenth-century *Heiji kassen ekotoba* illustration of the commander Minamoto Yoshitomo. The horns were mainly for decoration, and were made of gold or copper, and heavily ornamented.

A primitive headgear called a *hatsumuri* protected the cheeks and forehead of some warriors who chose not to wear a helmet, but this proved to be of limited efficacy, as the vital regions of the throat were not protected. During the fourteenth century, however, as warfare became endemic, a throat guard called a *hō-ate* was developed, which led to a

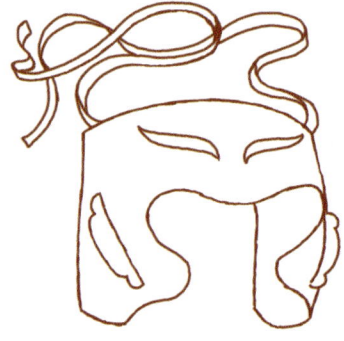

Hatsumuri *were favoured through the thirteenth century as a means of protecting the forehead and cheeks. The great weakness of this armour, however, is that it provided no protection for the chin and neck. In the fourteenth century, this style was replaced by* nodowa *and* hō-ate, *which protected the lower half of the face only.*

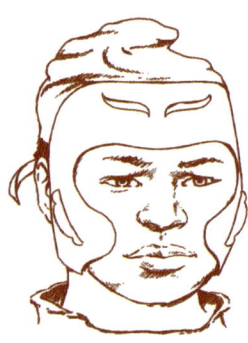

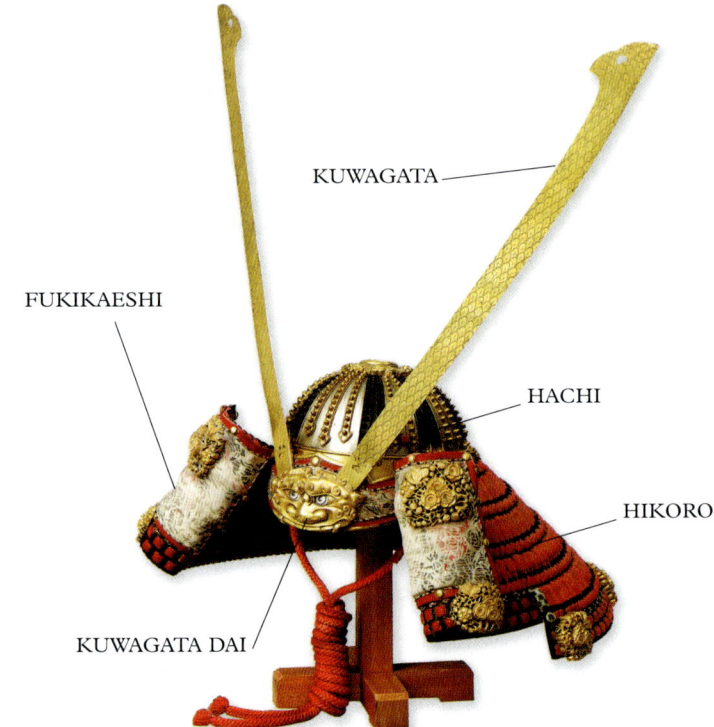

A helmet in the style of that of a twelfth- or thirteenth-century commander, with the prominent kuwagata, *and impressive decorations on the* kuwagata dai *and the* fukikaeshi. *Helmets with the largest* kuwagata *were often made as offerings for shrine gods. However, no commander would have worn a helmet with a* kuwagata *as long as depicted in this nineteenth-century copy.*

marked decrease in wounds to the throat. The existence of a *hō-ate* is also helpful in dating armour, because it came into existence only in the fourteenth century. A few warriors, such as the Miura, relied on full metal plate covering the head, cheeks and throat, and leaving only a small opening for the eyes and face, although this approach to using metal plates did not become popular. This innovation in armour led to a decrease in wounds to the face. This can be documented by counting the wounds mentioned in fourteenth-century petitions, submitted by warriors, that demanded rewards. Of all wounds suffered in 1333–38, 10 per cent were to the face, but by 1356–92 this figure had dropped to 2 per cent, revealing a marked improvement.

Another weakness of the helmet is that it provided little protection against concussions. One strategy in battle seems to have been to smash an enemy helmet as hard as possible with rocks or, from the fourteenth century, giant swords, for an enemy with a concussion or even one slightly groggy was easy to kill. Greater use of padded leather liners over time prevented warriors from being knocked unconscious on the battlefield.

ARMOUR AND CLIMATE

Armour provided relatively poor protection for the extremities. Over the course of the fourteenth century, 64 per cent of wounds were to the arms and legs. Gradually, however, armour became better at protecting the legs, but no notable improvement arose regarding arms. Armour did not initially protect warriors' legs, because there was no need for it to do so. Warriors initially fought on horseback had little need for extensive leg armour as their saddle and stirrups protected their legs. Some wore fur shoes (*tsuranuki*), which appear in the *Scrolls of the Mongol Invasions of Japan*, but these gave way to simple straw sandals in the fourteenth century. The straw sandal was favoured by foot soldiers as well, although some soldiers also went barefoot. Most had little, if any

SAMURAI HELMETS AND HEAD INJURIES

The samurai helmet protected all the head well save for the face, which remained vulnerable to a well-aimed arrow. Narratives of battle invariably describe how mortally wounded warriors were shot in the face or the neck. Any such injury would debilitate a warrior, for face wounds bleed profusely, and even a scratch to the forehead, or cheek will result in considerable blood loss. Picture scrolls reveal such warriors being escorted from the battlefield. Being shot in the eye entailed death or, if the arrowhead was only shallowly lodged, blindness. Shots to the throat and neck were more dangerous.

*Shin guards became more commonly used in the thirteenth and fourteenth centuries. They initially only protected the front of the shin. At times, they could be used as a rough and ready helmet. Takezaki Suenaga's shin guard helmet is shown here falling off his head. He also did not wear gauntlets (*kote*) to protect his hands.*

protection for their legs at all, save for simple robes. The protection of the saddle, stirrups and the low hanging flaps of armour proved adequate.

LIMB PROTECTION

Over the course of the fourteenth century, however, particularly after 1355, leg armour improved. Of all wounds suffered by the samurai from 1331–55, on average 37 per cent were to the leg, but during the last half of the fourteenth century, this figure dropped to 27 per cent. Improvements in leg armour account for the decrease. In particular, generals and wealthy warriors began to use chain metal or interwoven lacquered armour to protect the thighs in the fourteenth century. One image of this appears in a fourteenth-century portrait of the warrior Kō no Moroakira. Called *haidate*, this flexible protection for the legs became widely used in the fifteenth and sixteenth centuries, and proved valuable in protecting the upper leg.

Shin guards (*sune ate*) proved far more common and widespread than chain links. They are visible as early as the thirteenth century, but witnessed a great improvement in that they covered the back part of the calf,

MOUNTED SAMURAI

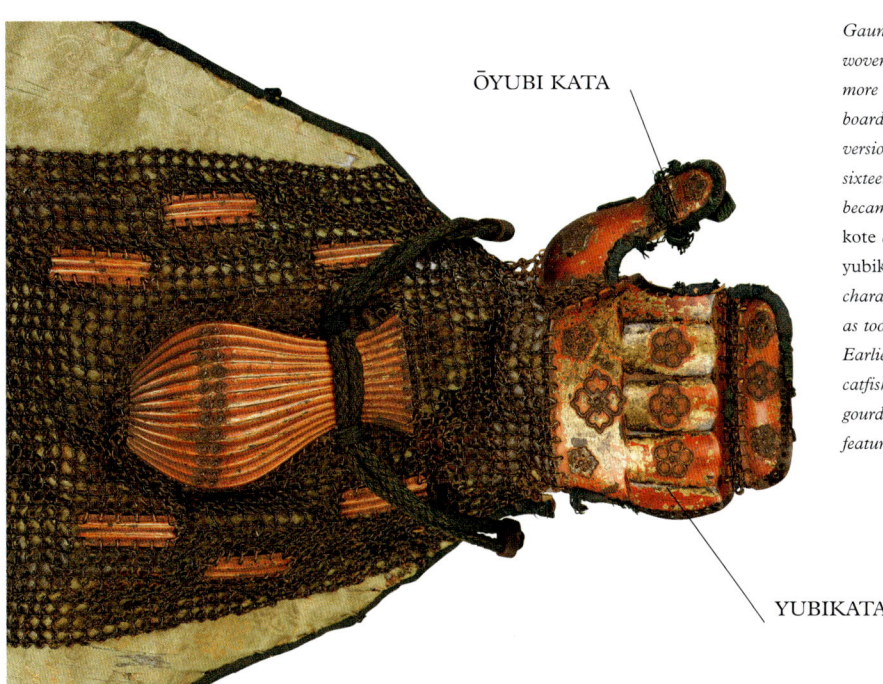

ŌYUBI KATA

YUBIKATA

Gauntlets (kote) consisted of iron chain links woven on fabric. These became necessary and more elaborate as armour simplified and shoulder boards were less likely to be used. The oldest version only protected the upper hand, but later sixteenth-century versions protected the arms and became increasingly elaborate, such as the Oda kote example depicted to the left. Note the yubikata, or finger patterns, which is characteristic of late sixteenth-century examples, as too is the ōyubi kata to protect the thumb. Earlier ones were described as looking like a catfish head and lacked this detail. The gold gourd-like decoration (fukube) is a defining feature of the Oda style of gauntlets.

GAUNTLETS

The *yoshitsune gote* most closely resembles older styles, with a simple *tekōgane* shaped like a catfish head covering the hand. This was attached with chords around the wrist and a loop for the middle finger. More elaborate late sixteenth century styles of gauntlets protect the lower and upper arm, and have much more elaborate ornamentation than the earlier examples. Here are depicted a variety of sixteenth century styles.

KOTE (LEFT)

KOTE (RIGHT)

YOSHITSUNE-GOTE

TSUTSU-GOTE

SHINO-GOTE

BISHAMON-GOTE

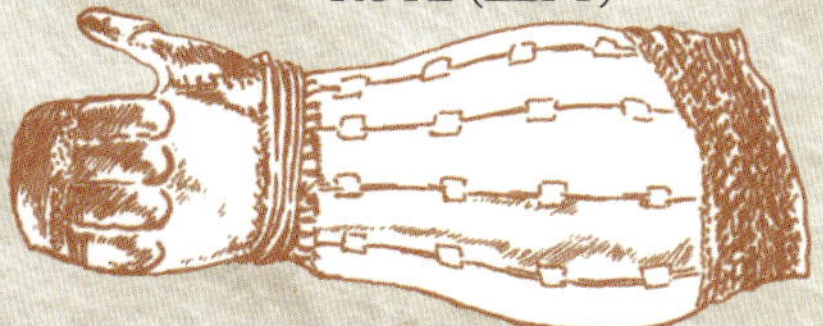

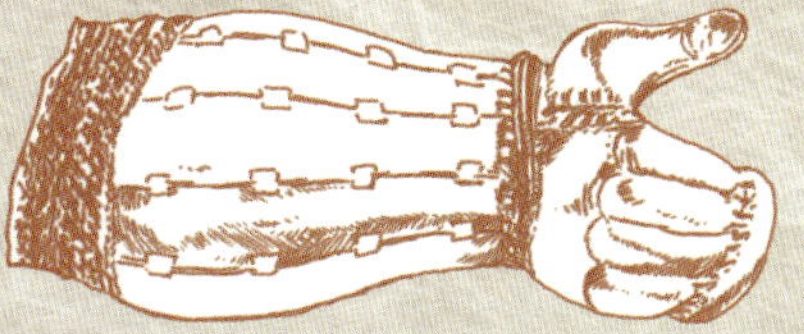

ABOVE: *The war boats of Shōni Tsunesuke (above) and Shimazu Hisachika (below), fighting against the Mongols. Note the platforms for rowers on the edge of these ships, as well as the flag of each commander displayed prominently. Only half of the warriors are wearing helmets with the rest preferring their* eboshi *caps.*

instead of merely the shin. They initially were composed of a narrow band protecting the shin, with two hinges on either side, which allowed this metal shin guard to protect the sides of the leg. As time passes, leg armour becomes more pronounced. In the fifteenth century, they came to be constructed from solid metal, with joints, protecting the front and back of both legs. The later shin guards could be made of a single piece of metal that was nearly round, or, more commonly, something with four hinges, to protect more thoroughly the front and back of the leg.

Arrow wounds to the arms and hands left warriors surprisingly nonplussed, which is why armour to the extremities remained relatively minimal. Padded cloth with chain linking, and small iron plating, called *kote*, protected the back of the hand and arm. One finds numerous examples of warriors being shot in the arm, hand or fingers and returning to battle just a few days later. Some archers would only wear these chain-linked *kote* armour on the left (*yunde*), or bow hand. Likewise, if they decided only to wear one sleeve, or *sode*, it would be on the left side, again, where they were most vulnerable when firing arrows. By contrast, foot soldiers, who might refrain from wearing *sode* boards on their upper arms, would often wear such chain link protection (*kote*) for their lower arms. The need to protect against arrows determined the style of Japanese armour,

Warriors waiting for a court case to be heard at Kamakura. The early samurai were among the most litigious people on earth, and were often involved in protracted legal disputes.

MOUNTED SAMURAI

in length. Despite their length, they could be fired on horseback because they were not held in the middle, but much closer to the bottom edge. The reason for this is not clear, save that it makes the wood of the bows less likely to break. The oldest surviving, and simplest bows were made from supple branches, or saplings (and known as *marugi yumi*), or strips taken from larger pieces of wood (*kiyumi*). Later, composite bows, whereby using glue made from deer innards bamboo was formed into the inner section of the bow (*fusetake yumi*). This bamboo proved less flexible, and allowed a bow, having been fired, to recoil quickly, thereby enhancing the power of its shot. These bows were enhanced even further, with long and flexible pieces of bamboo being glued to the outer edge (*sanmai uchi yumi*).

The use of bamboo meant that bow makers congregated in central and western Japan, where Japanese bamboo grows. The best bamboo came from central Japan, having a colder climate and in order to better understand why this was necessary, let us turn our attention to the bow.

MOUNTED ARCHERY

The bow constituted the dominant weapon of the samurai. The warriors of the thirteenth and fourteenth century referred to themselves as following the 'way of the warrior', which quite literally meant 'the way of the bow and arrow' (*kyūba no michi*). Implicit in this formulation was the notion that mounted archers shot their bows from horseback.

The bows of Japan are long, with some surviving examples, such as one bow in Kasuga shrine, being 187cm (6ft 2in) length, with the very longest reaching approximately 200cm (6ft 9in)

Minamoto Yoshitsune, a general who won the battle of Ichi-no-tani by charging with his ponies down a steep hill, was a short man with buck teeth. Later prints, such as the one here, depict him as a mighty commander on a powerful horse. The armour is in the style of ōyoroi, *but the shoulder boards (*sode*) and leg armour are depicted anachronistically.*

49

An image of a warrior from the Go-sannen ekotoba, *a picture scroll completed by 1347 that depicts the battles of Minamoto Yoshiie in the eleventh century. This scroll was prized by the Ashikaga shoguns, and is unusual for it attempts to accurately portray archaic styles of armour. Note, for example, the ōboshi* hachi *helmet depicted here. Early helmets were made from eight or ten plates connected with large rivets, which were thought to resemble stars, hence the name 'big star helmet'. By the time of the fourteenth century, far smaller rivets were used. Some helmets contained as many as 36 plates connected with 15 rivets each. Other elements of this illustration reflect, however, fourteenth-century norms. The long bow here is a compound bamboo type, which did not exist in the eleventh century.*

RIGHT: *Bow construction, cross section.* Kiyumi *(made from larger piece of wood, old type);* Fusetake yumi *(compound bow, with bamboo attached to the outside face of the wood using a paste made from fish or deer organs). During the thirteenth and fourteenth century, bamboo was added on both sides, making* sanmai uchi yumi. *In the fifteenth century, the wood was completely encased, producing* shihōchiku yumi.

KIYUMI

FUSETAKE YUMI

SANMAI UCHI YUMI

SHIHŌCHIKU YUMI

than the southwest of Kyushu, for it proved stronger than the bamboo grown in warmer climates. Bamboo was cut in the autumn (eighth lunar month) and, according to bow makers, the spring and autumn constituted the best time to glue composite bows together.

Arrows too had bamboo shafts, these being aged three years. The arrowheads themselves were made of iron or steel, and had a long shaft that fitted into the hollow bamboo. A variety of bird feathers were used to help the arrows fly far and true, with most being taken from birds of prey. Three or, at times, four feathers would be attached to the bamboo shaft. Eagle feathers proved the most distinctive, and were the most highly prized, hawk feathers could be used as well. The tail feathers of raptors were favoured, although wing feathers were another option. So important were these feathers that the samurai classified them in five categories, with the outermost tail feathers thought to be the most valuable. The feathers of other birds, such as the wing feathers of swans, were used, as well as the feathers of doves, but the feathers of owls, chickens and blue herons were never used. Some of the more decorative arrowheads, often given

RIGHT: Yumi *(bows). The one to the left is a lacquered wood bow, while the one to the right is a later style wood and bamboo composite bow.*

ARROW HEADS

A variety of tips were used. Some were reinforced steel, best used for penetrating armour, and known as war arrows (*soya*) while others, undoubtedly used to shoot at the face or neck, had two prongs (*karimata*) and were designed to inflict as horrible a wound as possible, and cause more damage when being extracted. Some two-pronged arrowheads had a turnip-shaped bulb attached behind it, immediately before the shaft. Known as a *kaburaya*, these arrows would emit a weird, low-pitched humming sound that often marked the beginning of hostilities. A variety of other styles of arrowheads became popular, as the varied examples depicted show.

HIKIME

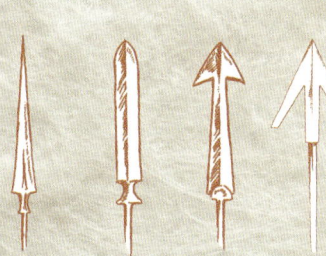
MAKU TOGARIYA TOBI-NAOSHI

KARIMATA

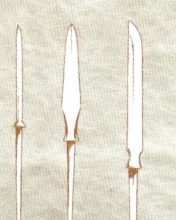
SOYA

ABOVE: Two types of Japanese arrow with flights. At top is a kaburaya, *or humming arrow, with a* karimata *tip; below is the more common* soya *– a short thin arrowhead of the type used in battle.*

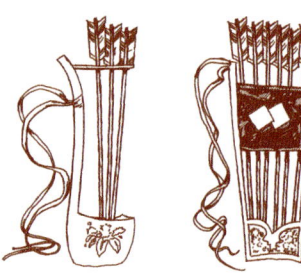

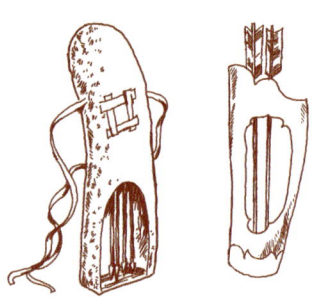

ABOVE: Types of quivers: ebira *(top);* yahoro *(bottom).*

to shrines, had a warrior's family crest emblazoned on them. Others proved to be rather primitive in construction, such as one two-pronged arrowhead in the author's collection. The arrowheads (minus the shaft) ranged in size from 2 to 4cm (0.8–1.6in) length, while the length of some examples with shaft included (again from Kasuga shrine) reached 10–11cm (4–4.3in).

On occasion, smiths would sign the shaft of an arrowhead – the author has one such example in his collection – but these objects never developed the mystique of the forged sword. The arrowhead shaft was inserted into the hollow bamboo, which comprised most of the arrow. The total length of some surviving specimens, such as one from Kasuga shrine, is 79.4cm (31.26in). Samurai kept some special arrows with their names attached, which they used only when aiming at high-ranking opponents, thereby allowing them to take credit for a kill. Arrows reserved for lower-ranking warriors did not, however, carry any sign of the archer's identity.

BOW STRENGTH

The strength of bows was measured by how many men it would take to string them, with some bows being described as three-, four- or even five-men bows. It is difficult to know how common these more powerful bows actually were, or whether the five-men bows even existed and were not literary hyperbole. Bowstrings were vulnerable to breaking, or could be damaged by rain. Warriors

FIRING A BOW

Unique among bows, Japanese bows are held not in the center, but from the lower half of the bow. The reason for this is unknown – perhaps it was because the oldest bows, made of trees, were less flexible and thus had to be held off centre, or perhaps it was because firing in this manner allowed for an archer to handle a larger bow. Firing this way, however, decreases strain on the hand and allows for more of a rebound, and thus a more powerfully fired arrow, than would otherwise be the case.

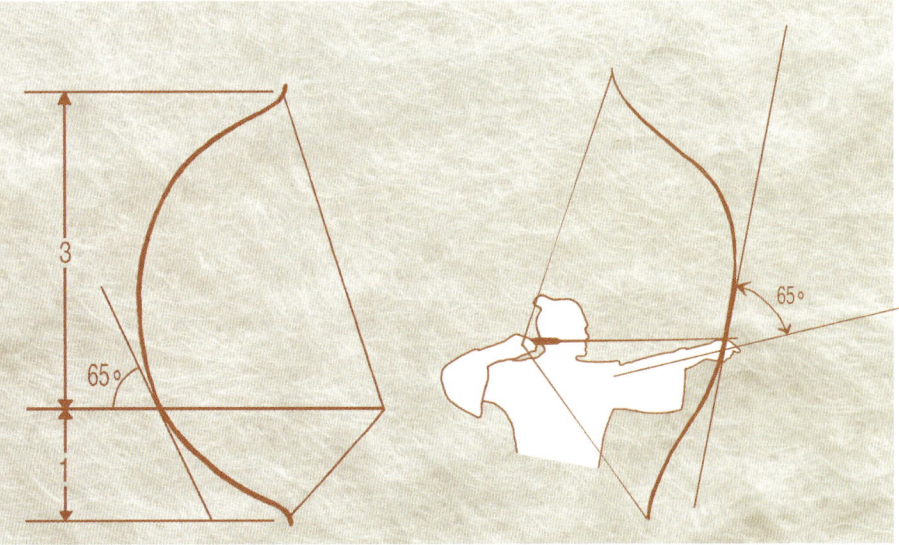

MOUNTED SAMURAI

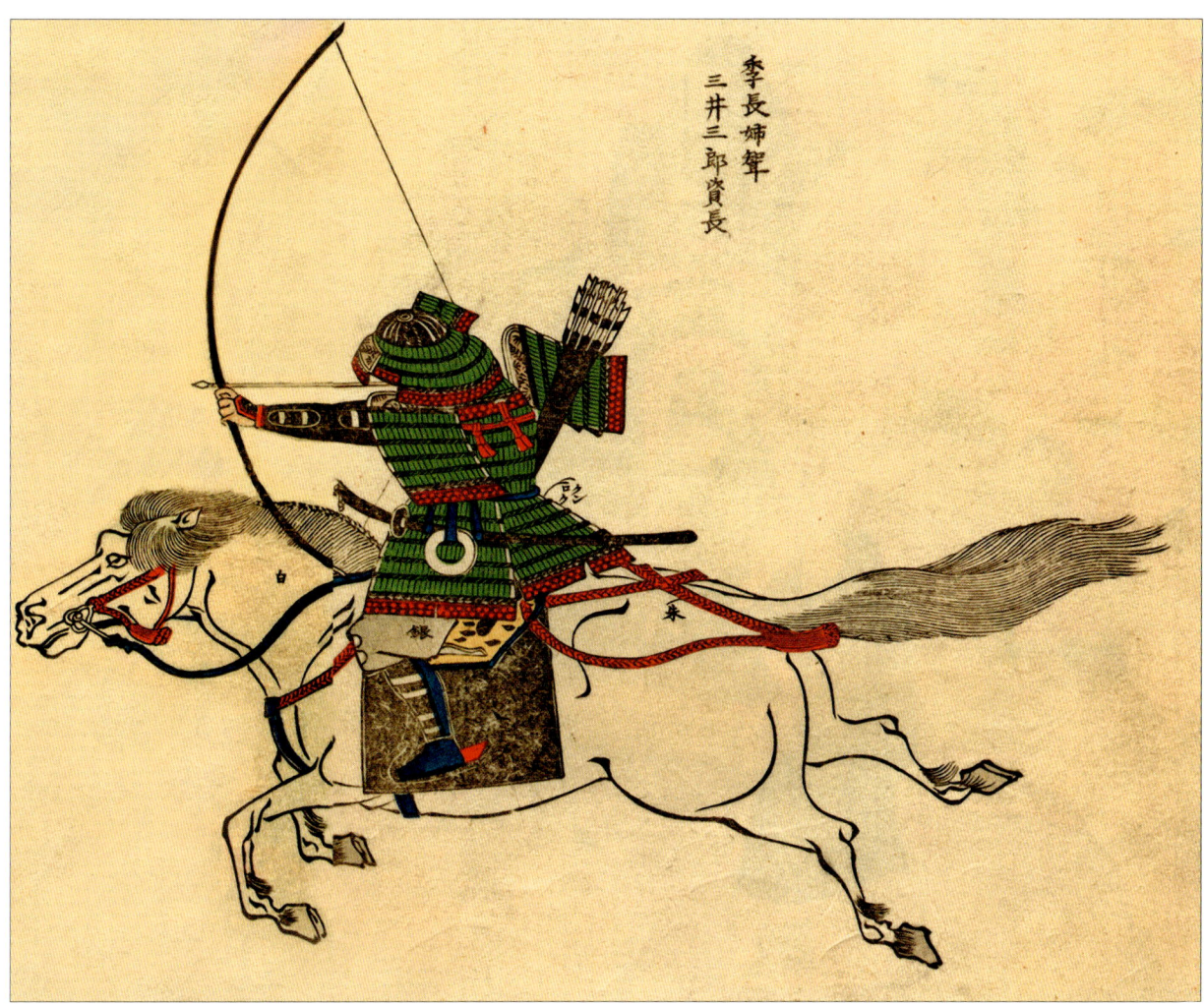

kept spares in a doughnut-shaped container called a *tsuruwa*.

Up to 20 arrows were kept in a square quiver. This quiver, in addition to holding a supply of arrows, would be used to store some food, such as a ball of rice, and *sake*, a favoured drink, particularly before the onset of battle. As time passed, however, the basket-shaped *ebira* gave way to a quiver that protected arrows with a fur cover, known as an *utsubo*, which can be seen in a famous portrait of the warrior Kō no Moroakira. Other warriors preferred instead a sack known as a *yahoro* to protect their arrows, one of which is depicted in the fifteenth-century *Yūki kassen emaki* picture scrolls. Mounted riders wore gloves to protect their hands when holding a horse's reins. Archers too required gloves, called *yugake*, particularly for their right hand, which was used to pull the bowstring. Reinforced leather protected the inside of the thumb and index finger, and indeed some gloves would only cover these fingers and the middle finger, which generally were used to hold the bowstring.

Swords were not often used on horseback, and so analysis of them will be continued in our next chapter, as too will an explanation of how mounted warriors fought with foot soldiers. Instead, in order to better understand the nature of battle, let us reconstruct

A warrior leading the charge, based on illustrations from the Scrolls of the Mongol Invasions. *Note how the reins are tied to the pommel board (*maewa*) of the saddle, allowing him to shoot with both hands. Samurai competed to be the first to engage the enemy, for the more one stood out on the battlefield, the more likely one was to receive rewards from one's superiors.*

a conflict waged by two forces of mounted samurai.

A CAVALRY ENCOUNTER

Mounted horsemen preferred open areas in which to fight. They particularly favoured relatively flat, dry terrain, although riverbeds remained popular places for horses to roam, particularly in heavily populated areas. Literary accounts have warriors announcing their names and fighting one-on-one against their adversaries, but short speeches between warriors seems to have been more of a literary convention to advance a narrative and identify characters, than a realistic option in battle. Indeed, identification remained so difficult that some warriors would designate an acquaintance as a witness. The warrior Takezaki Suenaga, for example, traded his helmet with the helmet of a comrade so as to be more readily identifiable to this man when he charged against the Mongols in 1274.

Even when fighting opponents from other countries, warriors wrote their names on their arrows. Kikuchi Takefusa, for example, shot a Mongol general in

SAMURAI, MONGOL INVASION

The Japanese archers established a reputation for accuracy. The invading Mongols apparently desired to add these warriors to their forces. Kikuchi Takefusa proved so skilled that he shot a Mongol commander in the face, thereby precipitating the recall of the first Mongol invasion in 1274.

the face in 1274, thereby precipitating the Mongol withdrawal. The Mongols were able to read his name, however, and their histories mention him as injuring their commander, an assertion that is matched by the Kikuchi genealogy.

Once two forces faced each other before battle, they would often yell out 'war cries', which consisted of a general or captain first yelling in a loud voice, 'Ei Ei' and his troops answering by bellowing 'Oh' with a crescendo. Each force would cry out, and the louder side, suggesting greater numbers, would have a psychological advantage. Thereupon a humming arrow, or two, would be unleashed and the opposing forces would advance.

These small forces were comprised of skilled horsemen and agile, if slow horses. They approached by cantering and would seek out an enemy who had advanced too far, or who was otherwise vulnerable. The conflict would entail great mobility and, at a key moment, a sudden charge. The charge was risky, however, for skilled archers, upon seeing an advancing enemy, would turn their horses away from the charging rival. This 'retreat' served to increase the amount of time an enemy had to close, thereby exhausting his mount. Furthermore, warrior preferred shooting behind their horses, or to the side, so as not to startle their own mounts. Warriors above all strove to conserve the strength of their horses. If a warrior could overtake an enemy whose mount had tired, and who had exhausted his supply of arrows, he could stab him from behind with a dagger. A few relied on long hooks, called bear's claws (*kumade*), which could be used to dismount a warrior from a greater distance.

Opposing groups of horse riders fought in close quarters, perhaps 20 or

ABOVE: *Japanese archers on boats fought much like they did on horseback. These small craft provided a platform for firing, little more. As the prominent quivers of these warriors reveal, most went to battle with around 16 arrows. Firing indiscriminately, or at a distance unlikely to harm an enemy, was unwise.*

The formidable power of a band of Japanese samurai is evident in this illustration from the Scrolls of the Mongol Invasions. Shiroishi Michiyasu and his men were able to break through Mongol lines. Scattered infantry could not withstand a charge by such a force.

30m (21 or 33 yards) apart, when they unleashed their arrows. As Japanese bows were very long, their arrows accelerated more slowly than those fired by shorter bows, for the kinetic energy stored in them was expended in releasing the longer bow itself.

They nevertheless could travel for long distances. Some accounts describe arrows exceeding 436m (476 yards), although the current record for firing an arrow from a Japanese bow is 385m (420 yards). Such long-distance skirmishes were called 'distant arrows' (*tōya*) and had minimal accuracy and killing power. Still, the stored energy of such a long bow allowed for heavier arrows weighing 50–70g (1.7–2.5oz) to be fired, which proved most effective in killing or wounding opponents in short ranges. Only at a relatively short range of 13–14m (14–15 yards) could arrows wound an opponent or penetrate his armour. At such ranges, Japanese bows can pierce even Teflon pans, meaning that iron-enforced breastplates were vulnerable. Likewise, one needed to approach an enemy fairly closely to be able to accurately shoot at his face, the most vulnerable area of the body.

In order to attack an enemy, a rider had to approach him from behind, either following a parallel path and overtaking him, or conversely, approaching at an angle from the rear, and fire his arrows to the side. The pursued warrior would then turn his mount to the side so as to be able to shoot back, and one can imagine the pacing ponies veering hither and thither.

A relatively cohesive group maintained an advantage, for its riders could shoot in many directions, but in order to attack and kill a rival, skilled horsemen would have peeled off in pursuit of stumbling enemies, or those who had unwisely tired out their mounts.

Likewise, injured or unhorsed riders were particularly vulnerable. An unhorsed rider found himself in mortal danger, for he could be surrounded and peppered with arrows by his mounted adversaries. Those lucky enough to fall in a ravine, or otherwise inaccessible area, might be able to survive, as too would one whose comrades came to his aid. Warriors were willing to help comrades as long as it did not endanger themselves or their horses.

CAVALRY INJURIES

Because samurai were loath to lose their valuable mounts, they rarely charged within a few metres of enemies. Most of their beasts were wounded by arrows, which generally proved not to be fatal. An encounter with a long sword or a pike could prove lethal. Good data for horse wounds exists for the years 1333–38, and a sampling of 31 horses reveals that 61 per cent of all

The inspection process. A warrior provides two enemy heads, which were verifiable proof of battle service in the aftermath of the Mongol invasions. A general, sitting on his armour box, inspects the evidence, and questions the warrior, before a report is made and submitted to Kamakura. A scribe here is recording the particulars.

horses were wounded by arrows, while 35 per cent were slashed with swords, and the remaining 3 percent were wounded by pikes. Arrows rarely inflicted mortal wounds, for only three out of 14 horses shot with arrows perished. By contrast, more horses slashed by swords perished than survived (eight out of 15), while the only beast that was stabbed by a pike perished. The horse statistics belie the fact that nearly all fighting comprised of exchanges of arrows.

To shift to data of human wounds, on average 73 per cent of all wounds from the fourteenth century were caused by projectiles, mostly arrows, with a few more caused by rocks. Swords caused 25 per cent of all remaining wounds, and pikes a mere 2 per cent. Thus on foot and horseback, most warfare consisted of skirmishing, with most warriors unwilling to risk their lives or their mounts in combat. So prevalent were these attitudes that to die in battle meant that warriors were expected to accrue great rewards in order to compensate for their loss.

Save for those striving to assert their autonomy, and who were willing to take great risks, most warriors strove to appear outstanding in battle without engaging in hand to hand combat. They accentuated their service while minimizing personal danger. This even led some warriors to show up in camp, have their presence confirmed by a general and then leave without fighting at all. Hatano Kageuji, however, recounts a battle in 1336 and states that his position was so precarious that he thought that he might be forced to commit suicide, and this desperation alone was the equivalent of dying in battle, and worthy of reward.

To conclude, mounted samurai conducted skirmishes on horseback, which were characterized by considerable manoeuvring and low casualties. Elaborate armour protected warriors in most encounters, which consisted of an exchange of projectiles, and the expense of this armour, and horses, caused only the most desperate or ambitious warriors to risk their lives,

Minamoto Yoshitomo, the commander of a band of warriors in the Heiji scrolls, is depicted at the upper left of the group of horsemen in this image. He wears red armour, hatsumuri *and a helmet decorated with two* kuwagata *horns extending from prominent dragon ornamentation. Yoshitomo holds a bow and arrow in his left hand and his arrows have distinct black eagle feathers. This is one of the oldest reliable portraits of a warrior commander, who notably does not lead the procession, but remains in the middle of the group. Generals were too important to engage in combat. The armour styles are consistent with those of the late thirteenth century.*

and those of their mounts, by directly trading blows with the enemy.

In cases where their homelands were threatened by a rival, or a marauding army, the prerogatives of war required very different behaviour, for the need to occupy lands and defeat an enemy necessitated riskier tactics and greater casualties. This led to a greater use of hand-held weapons, and the increased importance of skirmishers fighting on foot.

The Skirmishers

THE OUTBREAK OF CIVIL WARS NECESSITATED THAT THE SAMURAI TAKE GREATER RISKS. SOME DID SO WILLINGLY, FOR THOSE WHO WERE WOUNDED IN THE FIRST BATTLES OF A CIVIL WAR, AND COULD BOAST OF THEIR SERVICE, COULD EXPECT LAVISH REWARDS. BUT FOR EVERY WARRIOR WHO FOUGHT WITH EXTREME VALOUR, THERE WERE OTHERS WHO STROVE TO BE CAREFUL AND ENSURE THAT THEIR FAMILY WOULD PROSPER REGARDLESS OF WHICH FORCE ACHIEVED ULTIMATE VICTORY.

Detail of a group of samurai archers from the Heiji monogatari emaki.

A sword polisher. Polishing required great care, for otherwise a blade could be damaged, or its temper line obscured, thereby ruining its value.

A statistical comparison of wounds from the two major wars of the period, the Genkō and Kenmu Disturbance of 1333–38, and the Ōnin War of 1467–77, reveals common patterns. Hand-to-hand combat increased, with shock weapons (primarily swords and the occasional pike) inflicting 30 per cent more samurai wounds during the years 1333–38 than on average for 1339–92, while the years 1467–77 witnessed a 40 per cent increase in pike and sword wounds compared to what was normal for the rest of the fifteenth and sixteenth centuries. (For the purposes of this analysis, pike and sword wounds are treated as a single contrasting category from projectiles.) These periods also witnessed a greater level of fatalities than other times. Sixty per cent of all fatalities recorded for the years 1333–92 occurred during the years 1333–38.

These statistics should not lead one to assume that most wounds were caused by bladed weapons, such as the sword, for they accounted for slightly more than one third of all wounds (35 per cent in both 1333–38 and 1467–77) during their peak periods. In less intense conflicts,

HARAMAKI ARMOUR

FRONT VIEW

Haramaki was a simplified type of armour that was less roomy and box like than that of the 'great armour' (*ōyoroi*) of the mounted archers, and cost only one quarter of the price. The main difference in these two styles of armour is that the *waidate*, a separate board attached under the right arm, was abandoned, and instead, the armour was fastened together by tying the overlapping edges together. Sleeves were optional for *haramaki*, but often were attached by warriors.

Hirosode, flexible sleeves, protected the upper arm. They were designed more to protect the upper arm from blows than to function as portable shields, as the earlier sleeves had done, and came in use during the fourteenth and fifteenth century. *Sane*, small laquered reinforced plates, continued to be woven together by coloured cloth through the fifteenth century. *Katakami* were the chords used to hang this cuirass armor from the shoulders. *Kusazuri* protected the upper legs.

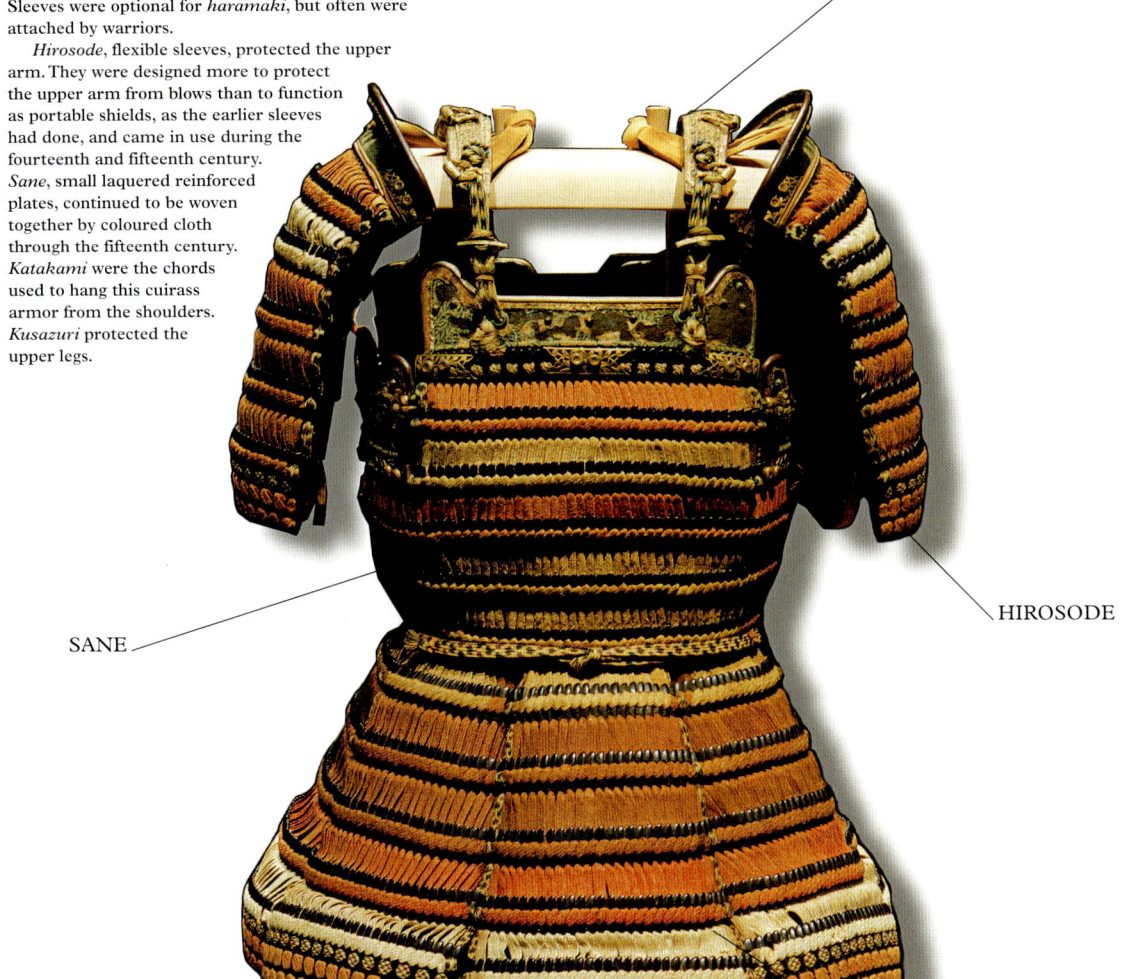

KATAKAMI

HIROSODE

SANE

KUSAZURI

THE SKIRMISHERS

such weapons caused only roughly one quarter of all wounds (27 per cent in the fourteenth century and 25 per cent after the Ōnin conflict). And even though earlier accounts only describe armies in terms of numbers of horsemen, by the mid fourteenth century one can find accounts revealing that at least one military unit consisted of 60 mounted warriors and 50 on foot.

We have already noted in Chapter 1, however, that arrows inflicted anywhere from two-thirds to three-quarters of all wounds inflicted during Japanese civil wars. Opportunistic skirmishing provided, in aggregate terms, perhaps an even greater impact on battles than the more dramatic duelling of horsemen, for skirmishers (*nobushi*) safely ensconced on inaccessible hillsides, or hiding on the roofs of dwellings, could unleash a hail of arrows, and unpleasantly surprise a nearby horseman.

Skirmishers consisted of both mounted and unmounted archers. We have already explored the fighting methods of the mounted archer. Let us now turn our attention to the skirmisher on foot. Skirmishes relied on their archery skills, and were more lightly armoured than the mounted samurai. Instead of wearing 'great armour' they preferred versions known as *dōmaru* or *haramaki*, two types of armour generally without shoulder boards.

SIMPLIFIED ARMOUR
The simplified type of armour called *haramaki* appeared at roughly the same time as *ōyoroi*. This armour lacked a supplementary section, called a *waidate*, which was attached to the torso section of great armour, and instead was composed of a single piece of armour that closed on the right side. Simpler in construction, and tightly fitting, this

OPPOSITE: *A portrait of Hosokawa Sumitomo (1489–1520) provides an excellent illustration of armour of his era. He wears* haramaki, *which can be seen tied under his right shoulder. Although in the style of older armour, this suit is typical for a general of his day. His* sode *are flexible, unlike the earlier versions, and his helmet, called an* akoda bachi, *is decorated with* maidate *horns attached to the front instead of* kuwagata *horns. A* nodowa *protects his neck. Sumitomo holds a long sword (*ōdachi*) with a one-metre (3ft) long wrapped handle, that was popular in his day, called a* nagamaki. *His gauntlets are more decorated than earlier versions, and the crest of Ashikaga shoguns and their favoured generals appears on them, as well as on his leg armour and horse accoutrements. His leg protection is far superior to earlier versions of shin guards. Called 'standing shin guards' (*ōtate age sune ate*) this solid metal leg armour protects the lower thigh, knee, shin and calf. Finally, the haunch of his horse is branded with a* mitsumeiyui *crest, composed of three linked diamonds.*

HARAMAKI AND HARA-ATE ARMOUR

Haramaki armour (below left) covered the body and was joined by overlapping panels underneath the right side. It provided more durable than two other common types of armour, that of the *hara-ate*, depicted here, which merely protected the chest and abdomen, and *dōmaru*, an extended version of the *hara-ate*, with increased protection for the sides, back and legs. Because *dōmaru* armour was connected simply, with no overlapping panels, it provided less protection than *haramaki* cuirasses. Strangely enough, *haramaki* and *dōmaru* armour types became confused in the sixteenth century, leading to the modern use of the terms to be diametrically opposed to the medieval usage, which is followed here.

Hara-ate armour (below right) consisted of simplified armour, comprised of *sane* woven together to protect the stomach and abdomen. This armour, hung from the shoulders, did not protect the sides or the back and was worn by only the lowest ranking fighters.

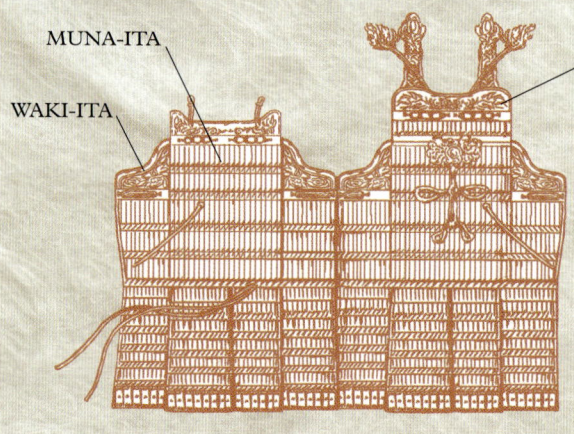

MUNA-ITA

WAKI-ITA

OSHITSUKE NO ITA

THE SKIRMISHERS

armour cost only one quarter the price of the more expensive great armour. Some frugal warriors favoured this armour, and added *sode* and other accoutrements, such as shin guards and metal gloves (*suneate* and *kote*), to make this armour the functional, albeit slightly inferior, equivalent of *ōyoroi*. Such an economical suit of armour would ideally be used for a *gokenin*'s *samurai* followers, while the highest-ranking warriors continued using the more expensive, and roomier *ōyoroi* armour.

Warriors fighting on foot, however, had no need for shoulder boards, instead preferring a simpler and even cheaper armour. The most basic armour, known as a *hara-ate* simply represented a reinforced piece protecting the torso. This armour proved very reasonable, but had obvious limitations in that it did not protect the sides or back at all.

In the early fourteenth century a new type of armour arose that was based on the *hara-ate*, but allowed for greater protection. This armour, called *dōmaru*, or 'round torso' had a lacquered enforced torso guard. Some of the earliest suits, such as one located at the Ōyama zumi shrine, still have their own shoulder boards attached. These oldest versions are also constructed by weaving together small platelets (*kosane*), meaning that their initial innovation was merely that they tied at the back and not to the side. Nevertheless, it lacks the ornamentation of the great armour, or for that matter, *haramaki*, and over the course of the fourteenth century, the use of shoulder boards became increasingly rare. Of course, individual preferences remained. Some full suits of *dōmaru* exist, complete with shoulder boards, but made entirely of hardened leather. Warriors of more limited means preferred such armour. One even finds a surviving suit of *dōmaru* that was tailored to fit the female anatomy and, according to lore, was used by a woman named Tsuruhime in 1542, during battles fought on Japan's inland sea.

It was acceptable for men of higher rank to wear *haramaki*, but *dōmaru* was reserved for warriors of lower rank, with the oldest documentary reference to such armour being a letter from 1271. Not all people fully understood the distinction between *haramaki* and *dōmaru*, and perversely, as time passed, many seem to have become more confused regarding the distinctions between them. From the sixteenth century onwards, armour tied at the back came to be called *haramaki*, while that tied on the side became known as *dōmaru*. The most important principle, however, remains that these styles of armour were better suited for soldiers on foot.

FIGHTING TECHNIQUES OF SKIRMISHERS

Skirmishers hid in dense undergrowth, or favoured muddy paddyfields or mountains. They congregated in areas where horses could not charge, and tried to shoot as many horsemen as possible. Wooden shields helped protect Japanese archers. These shields could not be held by a single hand, but instead were the size of large doors and were placed on the ground. If a group of samurai chose to charge on foot, one or two men would have to hold a shield and could not use any other weapons.

In more stationary encounters, shields were set on the ground, braced by a board to stand upright. Often, the doors of temples or other structures would be plundered for use as shields; these were of thick enough construction to stop almost all arrows. The fourteenth-century *Kasuga gongen kenki e* picture scroll provides an excellent depiction of how shields were used, with two opposing

NOBUSHI ARCHER

Skirmishers (*nobushi*), constituted warriors of all status, ranging from high ranking samurai to their humblest followers. They were scattered over the battlefield, and would generally fire at close range, although sometimes they would shoot 'distant arrows' for several hundred metres.

MUNA ITA

KUSAZURI

SANDALS

USING THE KATANA

BASIC DRAW

Japanese swords could be mounted so that the blade side was up, or down, in a scabbard. When placed down, and hung from the belt, the sword was known as a *tachi*, while when thrust into the *obi*, or belt, with the blade side up, the sword is called a *katana*. Originally, *tachi* were the favoured weapons, while lower-ranking samurai relied on *katana*, but over the course of the fourteenth century, the *katana* style became more popular, as it could be drawn more quickly. Here one can see a warrior in robes drawing his *katana*.

GUARD POSTURE

With a drawn *katana*, a samurai in guard posture could keep an opponent at bay. Holding a sword like this, he could easily swing the sword from the left, right or above, thus keeping his opponent off guard. Postures (1) and (2) show a low guard, while (3) and (4) show a higher guard.

BLADE DISTINCTIONS

There are a myriad of fine distinctions between blades. Some were composed simply of folded-over steel of uniform hardness, known as *maru kitae*, while other examples had a hardened cutting edge hammered onto a softer blade (*wariha kitae*). More sophisticated methods had a soft core surrounded by hardened steel on three sides (*makuri kitae* and *hosammai kitae*) while the final method had a soft core and hardened steel on all sides of the blade (*shihō-zume kitae*). In spite of these variations, the back edge of the sword was invariably thicker, while the cutting edge, demarcated with a temper line, proved harder.

All Japanese swords from the thirteenth century onwards are delicately curved. One great myth concerning the sword is that they were curved to better help warriors slash from horseback. In fact, the gradual and distinctive curve stems more from the process of tempering a blade of uneven thickness. Placing the red-hot steel in cold water caused the thin areas to contract more, leading to a distinctive curved shape. By contrast, swords of the earlier Heian era (794–1185) remained mostly straight, with just a slight curve by the handle (*warabitō*), with this difference stemming from characteristics of the metal used rather than aesthetic preference.

MARU KITAE

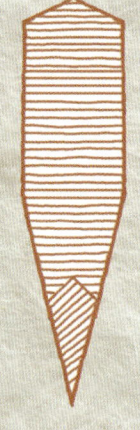
WARIHA KITAE

MAKURI KITAE

HOSAMMAI KITAE

SHIHŌ-ZUME KITAE

*LEFT: Drawing the sword is known as 'cutting the carp's mouth' (*kokuchi*) in Japan, because the entrance to the sheath is said to have a mouth like a carp. The sword is drawn by pushing the thumb of the resting hand against the hand guard, thereby releasing the sword for a swift draw. This is a highly skilled manoeuvre, as an error in technique could lead to the thumb being severely cut.*

groups firing from behind shields, in each case presenting a solid wall against the enemy.

Skirmishers did not solely consist of peasants; instead samurai joined their ranks as well, although it seems less likely that higher-ranking warriors participated. But in attacking across mountainous or difficult terrain, nearly all warriors dismounted and fought on foot. As time passed, skirmishing on foot became more common, which explains why improvement in leg armour occurred. The advantages of *ōyoroi*, namely its roominess, proved less advantageous for soldiers on foot, which is why simplified and leather armour became more common. Leggings protected the calves, shin and thigh, and were generally made from steel. Most foot soldiers remained skirmishers, as almost all battles were fought with projectiles. However, some preferred hand-held weapons, namely the sword, the glaive (*naginata*) and, in rare cases, the battle axe (*masakari*) and pike (*yari*). These weapons could only be used by the most skilled horseman and, even then, only rarely. Rather, these weapons were most effectively wielded on foot. Of all of them, swords were the most commonly used, and the most mystified, so let us analyze them first.

SWORDS

The sword epitomizes the samurai who arose in seventeenth-century Japan, when swords served as a marker of samurai

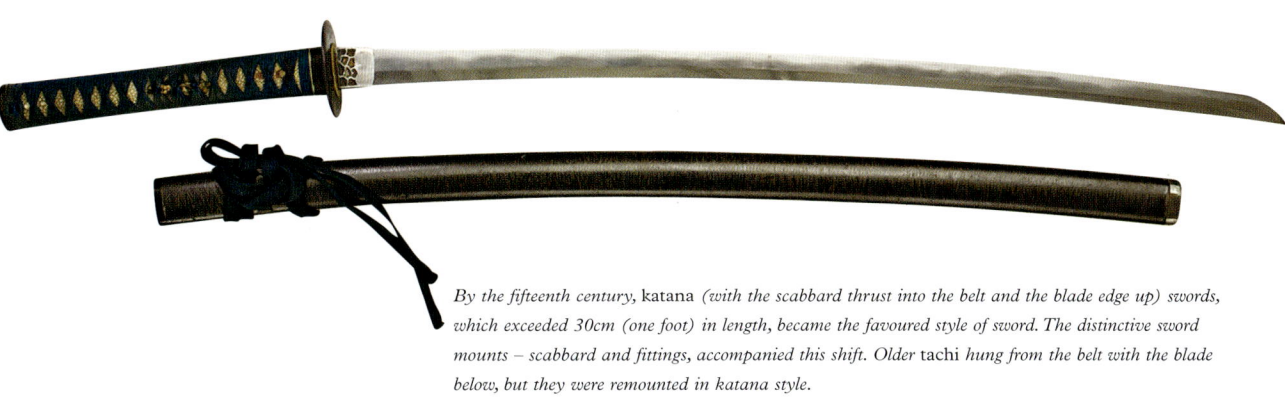

By the fifteenth century, katana *(with the scabbard thrust into the belt and the blade edge up) swords, which exceeded 30cm (one foot) in length, became the favoured style of sword. The distinctive sword mounts – scabbard and fittings, accompanied this shift. Older* tachi *hung from the belt with the blade below, but they were remounted in katana style.*

status. Prior to this time, they proved much less significant than bows. Swords were prized and valuable objects, although they were generally not as expensive as armour or a horse.

Unlike other weapons they were commonly named, and also bestowed to other warriors, or granted to shrines and temples for prayers, meaning that they survive in great abundance, particularly when compared to other types of weapons.

Swordmaking reached its apogee of manufacturing in the thirteenth century, when master craftsmen throughout Japan laboriously pounded and folded steel repeatedly so as to make the sword very dense, and at the same time flexible. Japanese swords had a hard cutting edge, and yet were less brittle than comparable weapons. They were not impervious to breaking, for one used with great force on an enemy helmet would most likely break. At other times, they were liable to be lodged in the body of an enemy, which explains why, if possible, warriors preferred stabbing an opponent to bludgeoning him.

Swords were not created with furnishings attached. Instead, they consisted only of a single piece of steel with a hole at one end, near which the maker's name could be inscribed. A bamboo peg could be inserted through this hole, so as to attach a wooden handle, coated in sharkskin and then generally wrapped with brown, or sometimes blue, cord. Small decorative objects, called *menuki*, representing anything from ears of rice to glowering demons, would be placed inside the handle. The edge of the handle, known as a *fuchi*, could also be similarly decorated. A *tsuba* or hand guard would protect the hand of the warrior, as well as providing additional space for creative decoration and embellishment.

Scabbards could be made of leather, but were most generally made of lacquered wood. They were often simple, but could be lavishly decorated. Inside the scabbard, a small knife was sometimes inserted, used as much as a glorified toothpick as for combat.

TSUBA – SWORD GUARDS

Iron sword guards were attached to a blade to protect the hand. The earliest ones tended to be relatively simple pieces of iron with some decorations carved into them. The centre hole is for the blade. An additional hole was for a *kozuka*, or small knife, which was housed in the scabbard. Later, the decorations became more elaborate and more open. A fine example of open work *tsuba* dating from the sixteenth century is illustrated to the right.

FOURTEENTH CENTURY

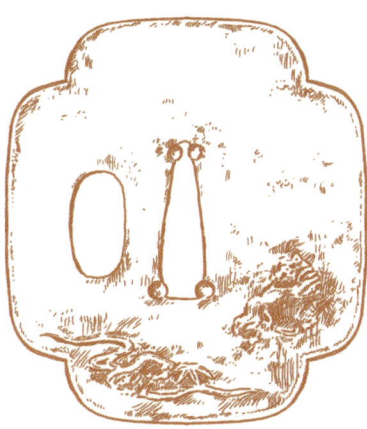

THIRTEENTH CENTURY

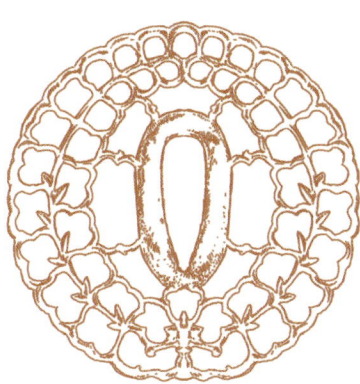

SIXTEENTH CENTURY

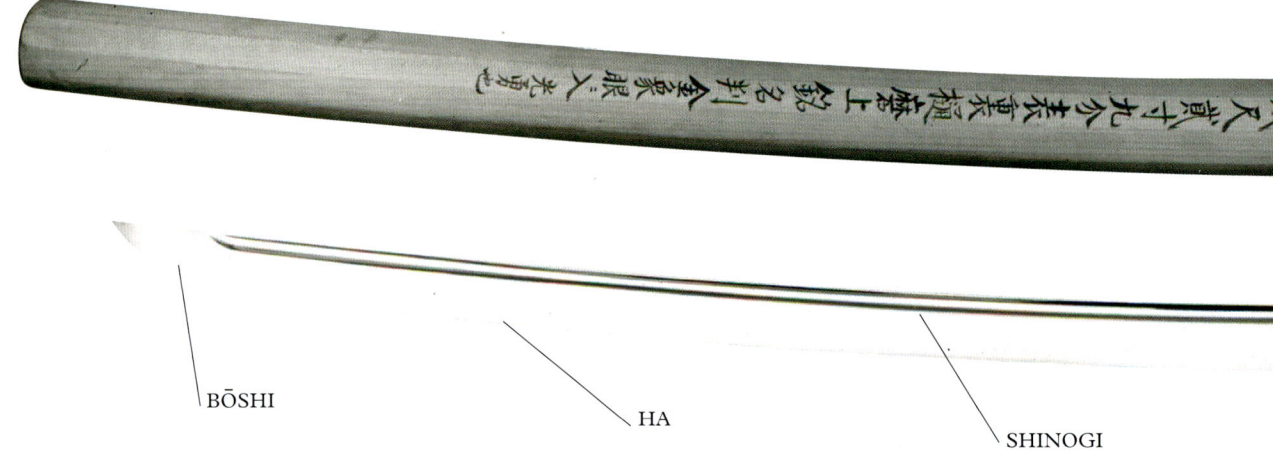

BŌSHI   HA   SHINOGI

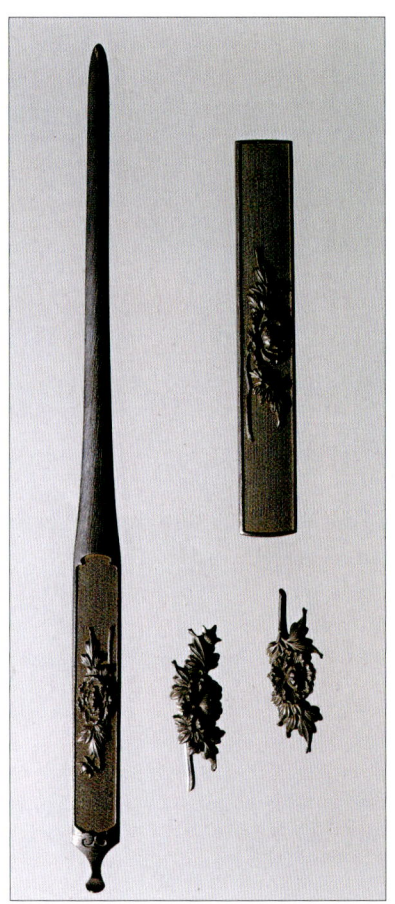

Fashions regarding sword fittings changed over time, as too did the most desired length of swords. Interest in very long swords, the *ōdachi*, peaked in the fourteenth century, with 2m (6ft 6in) specimens gaining popularity. As time passed, shorter swords were favoured, and often the sword was broken off in the handle region, and a new hole bored, to allow for smaller fittings and shorter scabbards to be used.

Swords were classified according to length, with the longer blades known most generally as a *tachi*. They could be worn with the cutting edge downwards, or, an approach increasingly favoured as time passed, with the cutting edge upwards, worn around a warrior's belt or kimono sash. The smallest knives, known as *tantō* or short blades, are generally under 30cm (11.8in), blades between 30 and 60cm (11.8 and 23.6in) are known as *wakizashi*, while swords longer than 60cm (23.6in) in length are known as *katana*. Slight variation exists, with the tips of swords known as *tachi* being more pointed and angular. They were gradually supplanted by the more rounded *katana*.

A set of sword fittings composed of a kōgai *skewer, the handle of a* kozuka *knife and two* menuki, *which were wrapped under the sword guard. These fittings all have the matching motif of peony flowers.*

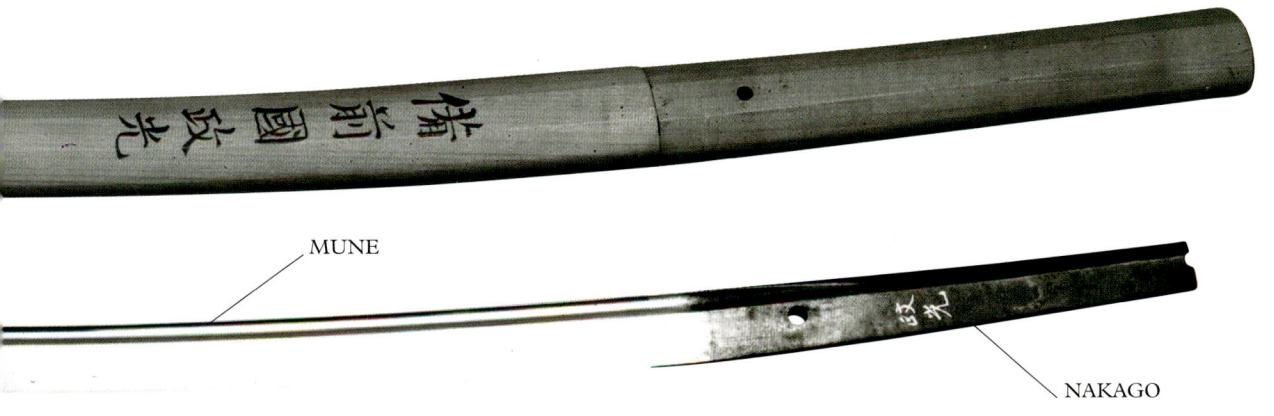

MUNE

NAKAGO

Swords were not invariably used to slash opponents. Some heavy unsharpened blades were used to smash an opponent's helmet, thereby rendering him unconscious and so helpless as to be dispatched by the samurai's followers. When the city of Kamakura was sacked in 1333, many people were killed and hastily buried at Zaimokuza. Slash wounds appear most commonly on the forehead or the top of the head, although they also appeared on the extremities of the skeletons of men, women and children. Some skulls were crushed due to the strength of the blows, while others display parallel cuts, revealing that the sword bounced off the skull, leaving several parallel strokes.

Some records reveal that men and horses endured multiple sword cuts and survived. One horse was slashed seven

ABOVE: A tachi *by the Bizen smith Masamitsu, c.1370. This photo of the sword shows it protected in a wooden scabbard. Valued swords could be stored thus and later placed in* tachi- *or* katana-*style fittings.*

HAND GRIP

For a strong grip, the *katana* is usually held at an angle of between 25 and 65 degrees with both hands. This allows flexibility of movement both upwards and downwards – by slightly flexing the wrist, the blade can be manipulated into a variety of attacking or defensive postures.

A wakizashi, *a short sword with standard fittings, a scabbard and a kozuka knife. Earlier examples could be placed in the belt, or in the robes, and were for personal protection, but in the sixteenth century, the* wakizashi *became popular as a short sword capable of being wielded with two hands.*

ATTACKING WITH THE KATANA

BASIC SWING

Katana were designed, above all, for cutting or slashing. The basic swing involves the sword being held over the head, and then with arms fully extended, involves a snapping of the wrists until ultimately the sword is held perpendicular to the swordsman's body.

ANGLE OF SWING

Katanas are most effectively swung with the blade straight up and centred over the head. Having it at an angle decreases the accuracy and power of the cutting blow.

CORRECT ANGLE

WRONG ANGLE

DIAGONAL SWING

A diagonal swing has the advantage of surprise, for until the last moment, an opponent would probably expect a 'basic swing' and the greater arc allows for an even more powerful blow.

LEFT AND RIGHT SWING

Katanas can be effectively swung from the left and the right. Note how the right hand pushes the sword forward when swinging to the right, while conversely, is used to pull the blade over during a swing to the left. Some medieval swordsmen were left handed, and thus held their sword in an unorthodox manner, which generally worked to their benefit, as their strikes would come from an unexpected direction or angle.

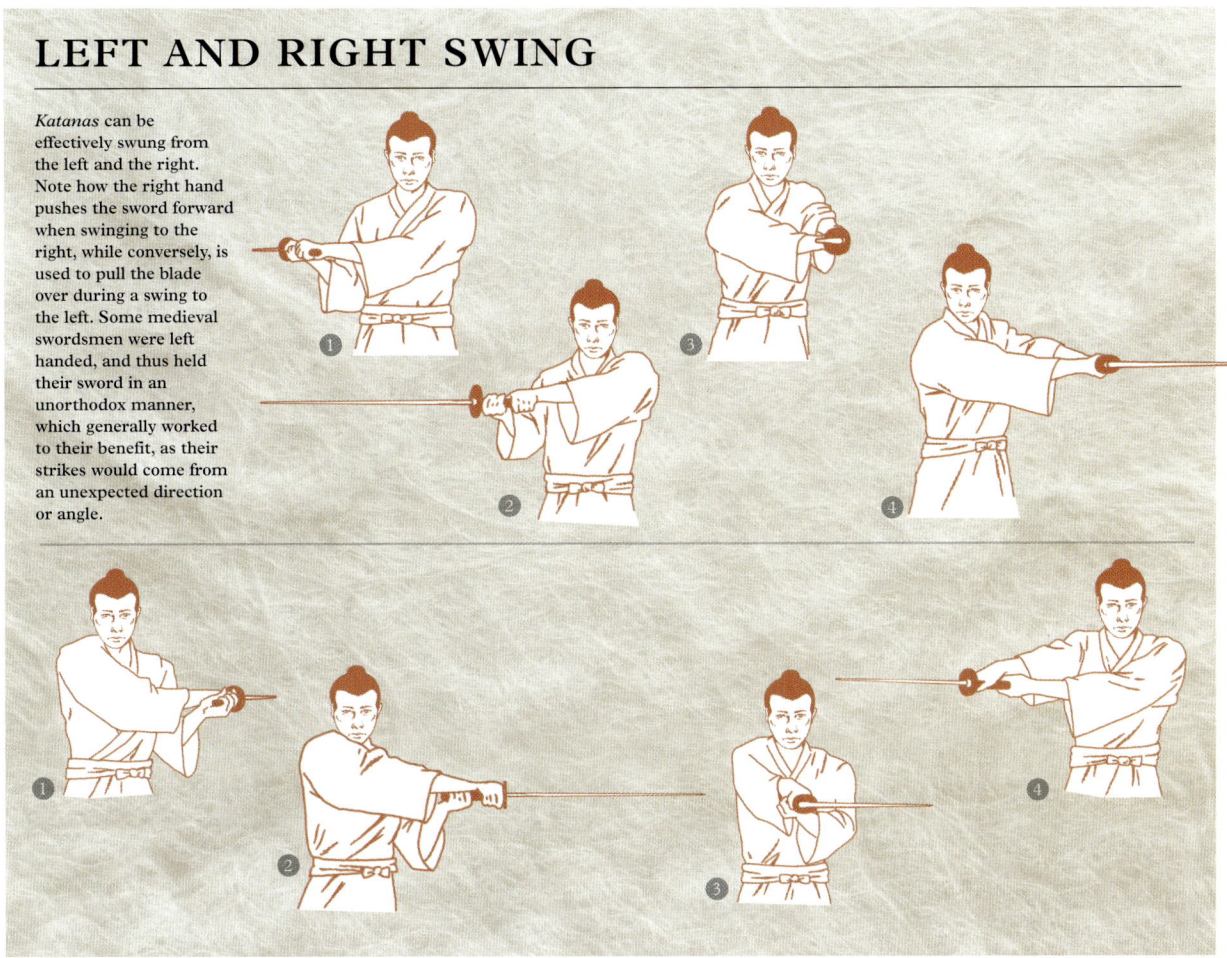

times, while another man was sliced 13 times – all of these wounds appear to have been minor, for both beast and warrior survived. The nature of these wounds also reveals that in combat those with swords approached just near enough to a rival to nick them with the tip of their blades. Samurai did not readily approach an armed opponent, even when they resorted to hand-to-hand combat, instead they remained as far apart as possible, save to inflict the most glancing of blows. An incapacitated warrior, or one unaware of his or her rival's presence, was more likely to suffer a crushing blow.

Old swords, particularly of the twelfth and thirteenth centuries, do not seem to have been sturdy enough to pummel an opponent, but the thinness of their blades often stems from repeated sharpening through the centuries. Earlier blades were a comparable width to later ones, although during the fourteenth century some unusually large swords gained favour.

The fourteenth century witnessed the innovation of the *ōdachi* blades, which could attain a length of 2.1m (7ft). These weapons proved remarkably well balanced, and surprisingly easy to wield in spite of their length. No scabbard was generally used, and instead warriors held them on their shoulders. Such weapons likewise could not be wielded on horseback. Instead, they provide a unique insight into the nature of fighting in the fourteenth century.

The great sword was sharpened at the tip but not on the edge, which was characterized as a 'clam shell' blade (*hamaguri*). This reveals that stabbing and bludgeoning were the primary uses of this weapon. These swords could be used to break the legs of horses, and also to smash enemy opponents on the head. They had the advantage of other similar bladed weapons in that they were more durable, because they lacked a wooden handle, although still ones sees example of even giant swords breaking in combat – steel helmets were unforgiving for even the best swords in the world.

INDIVIDUAL COMBAT

The fact that long swords were used reveals, as well, that warriors on foot did not fight in cohesive formations. Instead, each warrior would swing or stab his weapons at all who approached

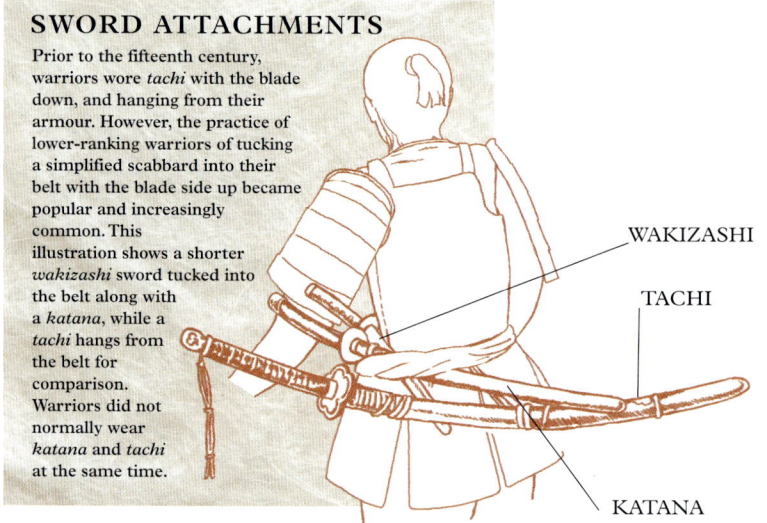

SWORD ATTACHMENTS

Prior to the fifteenth century, warriors wore *tachi* with the blade down, and hanging from their armour. However, the practice of lower-ranking warriors of tucking a simplified scabbard into their belt with the blade side up became popular and increasingly common. This illustration shows a shorter *wakizashi* sword tucked into the belt along with a *katana*, while a *tachi* hangs from the belt for comparison. Warriors did not normally wear *katana* and *tachi* at the same time.

WAKIZASHI
TACHI
KATANA

him. None, neither friend nor foe, could approach within a 2m (6ft 6in) perimeter of such men. Instead, only two options were available to counterattack. One, the easier way, would be to lure or trap such a warrior in an open field, and then shoot him repeatedly with arrows. The Shimazu, excellent mounted archers, made short work of such foot soldiers in 1333 by wheeling around them, always staying out of range, and peppering them with arrows, until these most skilled swordsmen were dead. One member of this troop summed up such an encounter by stating 'even the strongest warriors cannot withstand the bite of arrows nor can the fastest of men outrun a horse.' Another, more desperate recourse, would be to time an attack just right, and tackle the warrior by closing the distance after he had swung his sword and stab him with a dagger; as we have seen, the edge of the long sword was not sharp and would thus be ineffective in close quarters.

CAUTIOUS COMBAT

Even the most skilled swordsman with the longest sword had to either hide in a protected area, or cower behind a door-like shield, and wait as squads of horsemen cantered by. Any unwary horsemen or unlucky rider thrown or dropped from his horse would become a target, but the sensible rider would not approach inaccessible or occupied terrain. A scene from the *Kasuga gongen kenki e* picture scrolls epitomizes how swordsmen fought cautiously, clustered in small groups, with shields to the fore, waiting as two others advanced to dispatch a wounded opponent.

Long swords could also be used to fight other foot soldiers. In this case, the person with the longest sword had the greatest advantage in striking range, although considerations of weight and unwieldiness meant that slightly over 2.1m (7ft) represented the greatest length of these blades. A scene from the late fourteenth-century *Aki no yo no nagamonogatari* scrolls ably reveals such an encounter, for it shows a group of warriors fighting on foot, with some shooting the others with arrows and others attempting to stab each other.

PIKES AND BATTLE AXES

Although the *ōdachi* long sword proves the longest and most durable weapon used during the fourteenth century, other hand-held weapons were used as well. Pikes (*yari*) were in documented use in 1334. They were only approximately 1.5m (5ft) in length and thus shorter than long swords. The oldest depiction of these weapons, appearing in the *Kasuga gongen kenki e*, reveals a fallen warrior wearing only simplified armour (*dōmaru*, with no shoulder boards or leg protection) with a pike underneath him. One can deduce from this scroll that pikes were composed of a short blade (*tantō*) inserted in what appears to be a piece of wood, or perhaps bamboo. These weapons were more inexpensive than great swords, but also relatively ineffectual, not being as long as the great sword, nor as durable, nor as able to slash an opponent. Hence, during the wars of the fourteenth century only 15 wounds, a mere 2 per cent of all recorded injuries, can be verified as being caused by pikes.

Battle axes were also favoured by some warriors, and one can find literary references to warriors using them, as well as visual representations dating from the mid fourteenth century, such as the *Go*

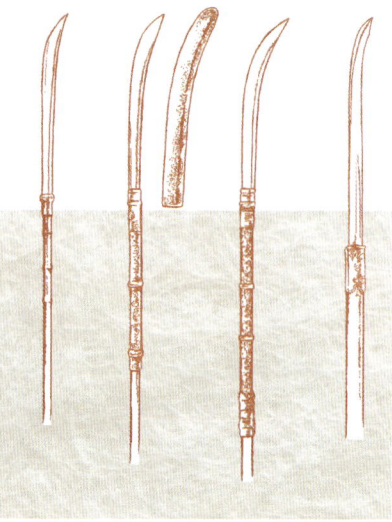

NAGINATA BLADE TYPES

Naginata are curved blades with a long tang that is inserted into a wooden pole. Variation existed, in that some where largely straight with a curved tip, while others were more generally curved. They could be used for both slashing and stabbing, and were a more versatile weapon than pikes.

THE SKIRMISHERS

HEISHI
(THIRTEENTH CENTURY)

Heishi, Low-ranking warriors, wearing minimal armour, such as this man, also participated in battle. He is wielding a *naginata*, has a small sword in his belt, and also wears minimal armour, in this case a *hara-ate* designed to protect the chest and torso. Instead of metal shin guards, he wears *habaki*, a cloth-like covering for his legs. The man here has straw sandals, which became increasingly popular, although some *heishi* remained barefoot.

HARA-ATE ARMOUR

NAGINATA

HABAKI

WARAJI

Sannen ekotoba picture scrolls. Akamatsu Ujinori, a lesser scion of an important fourteenth-century *shugo* family, used his battle axe to smash the helmets of several enemies until an adversary struck back, breaking his handle with a sword. A functionally similar weapon, and one far more common, the halberd or glaive (*naginata*) was used to either slash or stab. Both these bladed weapons were better at slashing, than merely stabbing like the pike, although their wooden handles meant that they were not as durable as long swords.

EARLY FORTIFICATIONS

The onset of civil war allowed skirmishers to have a greater impact on the battlefield, for mounted warriors were at times forced to fight in cities or in inaccessible terrain. In cases of urban warfare, which meant the capital of Kyoto, skirmishers hid behind structures or shot at horses from above, thereby forcing cavalry groups to either roam along riverbeds, or to burn areas that they needed to hold.

Those on foot desiring to hold an area needed to construct fortifications, which proved to be generally small and hastily assembled affairs. One of the best depictions of a fourteenth-century castle, the *Rokudō-e*, reveals how a small band of warriors occupying a hill had built up a wall with barricades, and in front of that stationed a number of archers behind heavy shields. To call such structure a castle is a misnomer. Castles were small structures, manned by only

A naginata *blade. The tang, which is as long as the blade, would be inserted into a wooden pole. Its great length ensured that the blade would be securely attached to the pole. The extra weight helped balance the weapon and prevent the pole from being easily broken. These durable weapons, suitable for stabbing and slashing, were cheaper than swords, and supplanted only when armies adopted cohesive units armed with pikes.*

BATTLE AXE

Battle axes (*masakari*) were a common weapon in the fourteenth century, used by *ashigaru* and other peasant infantry, but their use fell from favour in later centuries as spears and pikes became more effective battlefield weapons.

50 or 60 men, with the largest holding up to a hundred.

Many castles were constructed from dismantled dwellings and lumber. Some were used a mere ten days after their initial construction, and their effective lifespan was measured in a few weeks or months, and they only rarely lasted several years. Some castles represented

The remains of Taihō castle, an unusually elaborate and well constructed fourteenth-century castle. Its earthen defences, now shrouded in vegetation, proved sufficient to hold off a besieging army for months. This castle could be supplied by boat, for the lowland in the foreground constituted an extensive swamp.

fortified dwellings, while others corresponded with mountain temples. A few, such as Seki and Taihō castles in eastern Japan, were elaborately constructed structures, located in swampland and reinforced with earthwork walls. These castles represented a strategic core of the Southern Court's presence in eastern Japan. The Southern Court referred to Emperor Go-Daigo and his successors, who established their capital in Yoshino, to the south of the capital. In contrast to their bases in central and western Japan, the Southern Court attracted only limited support in eastern Japan. Their partisans fought against the more

numerous Ashikaga from 1336 until their surrender in 1343. Their castles proved solid enough to withstand saps constructed by the besieging army, but this durability was exceptional in the fourteenth century.

MOUNTAIN CASTLES

Mountain castles had the advantage of easily accessible large boulders that could be rolled onto advancing enemies, and one finds records of hapless bannermen having their heads crushed by boulders. Save for these simple devices, one does not find many examples of advanced siege engines, or anti-siege engines, such as the trebuchet, being used prior to the

THE SKIRMISHERS

WARRIOR MONK

These men were known as *akusō* or 'bad monks' who engaged in violence in order to protect their temple's interests. Violence was not stigmatized among the clergy as long as it was for a good purpose, such as upholding the interests of a temple or aiding political patrons. Young monks fought, while older monks employed curses, which were also thought to be acts of war. The monk depicted here wields a *naginata*, a weapon favoured by monks.

HARAMAKI ARMOUR

KATANA

at the top of a hill, but approaching the castle all paths are raised, and these funnelled advancing enemies along a narrow corridor, where a strategically placed foxhole allowed a handful of archers, well armed and supplied, to hold up a much larger enemy indefinitely. This castle, used in the fifteenth century, could hold out for half a year, for mountain paths linked to a nearby temple allowed for easy transportation of supplies that could not be readily cut off. The presence of such foxholes, the medieval equivalent of pillboxes, emphasizes the sophistication of these structures, for they allowed for offensive attacks within the larger defensive structure.

Each fortified structure represented a web of outlying and supporting fortified regions, often located on hilltops, each of which overlooked roadways, and could be readily supplied. This network aided in defence, and also ensured that no one structure proved of crucial importance.

Often a cluster of castles located only a few miles apart, on both hills, and flatland, determined control over an area.

In order to gain control of a heavily fortified region, an extended campaign was necessary, with a core of territory held and outlying structures gradually reduced one by one. More than five years of fairly constant battles were required before Seki and Taihō castles in Hitachi Province were captured. These structures were particularly difficult to take, as they were situated in swampland, heavily fortified and regularly reinforced by forces in small barges.

FOURTEENTH CENTURY CASTLES

Fourteenth-century castles, even the structures at Seki and Taihō, were not physically impressive, being built primarily from wood and earth. The nature of their construction meant that in order to defend such flimsy structures, all nearby wooden structures not appropriated for use in building castles might be burned down so as to prevent

mid fifteenth century. For castles to withstand a longer siege, they required considerable supplies, such as several hundred bales of rice, other grains such as barley, and beans and rice chaff (*nuka*) used for fodder. Access to water proved essential, as those castles that were rapidly constructed in areas without wells could be more easily captured.

Nevertheless, the capture of even the humblest of forts was not easy. Often, even a single castle could have two separate cores, which meant that part of the castle could fall, while the other one would remain untaken.

One castle, located in Higashiyama, the mountains to the east of the capital, only occupies some 30 sq. m (323 sq. ft)

ABOVE: Japan's warriors constructed extensive walls, first made of earth and then supplemented with stone, to defend Japanese harbours from the Mongols in the aftermath of the 1274 invasion attempt. These durable walls prevented the Mongols from landing in 1281 and some portions survive intact to this day.

BELOW: A scene from the late thirteenth-century Ippen hijiri e, *depicting a small station. Travellers and goods flowed easily throughout thirteenth-century Japan. Stations were ideal staging places for armies and necessary for the transportation of military supplies.*

enemies from using their wood or food for supplies. In cases where building materials were readily available, clusters of castles were built within a few miles of each other, and supported by a small number of men.

STONE DEFENCES

Not until the early decades of the sixteenth century was stone used to fortify weak spots in the defences. Stone walls were, however, constructed in 1275, and were designed to protect Japan's most vulnerable harbours in Kyūshū from a Mongol invasion. In this case, earthworks were initially constructed but thought to be inadequate, hence they were reinforced with stone, making them approximately 2–3m (6ft 6in–9ft 10in) high and of similar width, stretching for 10km (6.2 miles) along the beach. These walls proved effective for they prevented the Mongols from landing on the soil of all but a few peripheral islands for six weeks, before a typhoon destroyed their fleet in 1281. Samurai warriors themselves were responsible for construction of the walls

in 1275, with the burden determined by the amount of lands they possessed. No standardized means of construction existed, for the stone walls exhibited differences based upon the province where the warriors resided. Warriors relied on either granite or sandstone to construct the walls, and depending on the terrain and availability of rocks either used nearby materials or shipped them from as far as across the bay of Hakata or, for example, from the island of Nokonoshima. Likewise, the walls were kept in repair for decades, with the latest recording repairs occurring in 1338, over 60 years after the initial construction.

THE PITCHED BATTLE: AN OVERVIEW

Having provided an overview of fortifications, and of the dominant weapons on foot as well as the weapons of mounted samurai (see previous chapter), we can now analyze a specific battle, one that took place on the last day of the sixth month of 1336. This battle has been selected because it represents a decisive encounter that led to the establishment of the Ashikaga *bakufu*, Japan's second warrior government. In contrast to earlier conflicts against the Mongols, or for that matter the civil war of 1180–85, this campaign is well documented. Although the Ashikaga decisively won this battle, they were unable to annihilate their rivals.

Warriors who fought in battle demanded compensation for their actions. They submitted petitions recounting all damages suffered, or wounds incurred, and also listed where they fought. The survival of documents known as battle reports (*kassen chūmon*) and petitions for reward (*gunchūjō*) allows historians to reconstruct how war was waged, for these documents also record how wounds were inflicted. Such documents first appear late in the thirteenth century, in the aftermath of the Mongol invasions, and were

A bannerman advancing against the enemy, from the Mongol Scrolls.

standardized by the onset of the civil wars in 1333. The earlier records provide a narrative of a warrior's actions in battle, and approximately 1300 documents survive for the fourteenth century. These mention 8634 individuals, the locations where they fought and 721 cases where battle wounds are described in detail. Written shortly after every skirmish, each document also mentions the damages incurred by warriors so as to ensure compensation for their actions.

Failure to receive adequate remuneration meant that the warrior might fight for the opposing side, something that some families did with frequency. At times samurai would write that they would fight for the enemy 'unless they received sufficient rewards', an attitude that some commanders would castigate as being 'the attitude of merchants'.

The conditions in Japan in 1333 proved unsettling, for the country witnessed the sudden and violent

WEAPONS AND FIGHTING TECHNIQUES OF THE SAMURAI WARRIOR

A nineteenth-century woodblock print of Ashikaga Takauji (1305-1358), the founder of Japan's second warrior government (1338–1573), presiding at a council with his generals and men.

destruction of its first warrior government, the Kamakura *bakufu*. The emperor Go-Daigo (1288–1339), who strove to rule as an authoritarian and absolute ruler, plotted against and ultimately destroyed Kamakura, and his success stemmed largely from the defection of one of Kamakura's greatest generals, Ashikaga Takauji (1305–58). Flush with victory, Takauji desired to receive the title of shogun (a military ruler technically subordinate to the emperor), which in turn allowed him to assert the unilateral privilege of recourse to military force, but Go-Daigo demurred, for he wanted to rule as an absolute ruler who monopolized the use of military force. Outside events forced their hands, for a rebellion by surviving Kamakura supporters proved startlingly successful in the summer of 1335, and Ashikaga Takauji set out to quell this uprising without permission to do so. Takauji quashed the Kamakura uprising, and then occupied the ruins of Kamakura, where he started rewarding his followers and unilaterally claiming to be shogun. Instead of rewarding Takauji, Go-Daigo strove to chastise him for his insolence. Takauji, however, defeated Nitta Yoshisada (1301–38), the general dispatched to quell him, and advanced on the capital late in 1335, before briefly occupying it during the first month of 1336 until Go-Daigo's reinforcements from the north forced Takauji to flee to the west of Kyūshū, where he triumphed in the battle of Tadarahama and then rebuilt his forces. By the fifth month of 1336, he returned to the capital, and won a crushing victory, killing Go-Daigo's most skilled general, Kusunoki Masashige (1294–1336). He then occupied the capital, where he encamped at Tōji, a temple in the southern ward. Go-Daigo, by contrast, moved to Mt Hiei, a mountain to the north-east. Neither force could dislodge the other, and the battle in the capital continued until the end of the eighth month of 1336, when the Ashikaga ultimately drove Go-Daigo's forces from the capital and established a new regime, the second warrior government, or Ashikaga *bakufu*.

RECONSTRUCTING THE BATTLE FOR THE CAPITAL, 30 JUNE 1336

According to Ashikaga lore, the battles of 30 June 1336 marked the advent of Ashikaga hegemony. Opposition remained in the capital, but the Ashikaga no longer had to remain on the defensive, clinging precariously to an encampment in the Tōji temple grounds. This battle, like many decisive encounters, spread its share of legends and traditions. One eastern gate of Tōji, for example, remains closed to this day because Takauji himself did not go out in order to fight the waiting Nitta Yoshisada and his army. Likewise, the Ashikaga believed that they had divine support, for humming arrows peppered the Nitta forces from a shrine located in the temple precincts. Unlike many legendary battles, however, this encounter can be reconstructed in great detail from numerous documents submitted by warriors who fought for both sides.

The battle for Tōji represents the culmination of a three-week campaign for control of the capital. On the eighth day of the sixth month, the Ashikaga forces attempted to storm Mt Hiei, where Go-Daigo's forces were encamped. They killed one of his generals, Chigusa Tadaaki, as they climbed the narrow trails of this 848m (2782ft) mountain, but were ambushed

*A fourteenth-century portrait of Ashikaga Takauji, in court robes and with an imperial guardsman's sword (*efu no tachi*). The Ashikaga emphasized their court ranks and positions as much, if not more, than their warrior origins.*

with rocks and shot with arrows, suffering heavy casualties. One warrior, Katayama Takachika, for example, would mention how his bannerman had been shot in the right thigh, while one of his samurai, called a 'child of the house' (*ienoko*) in this document, was shot in the left hand and thigh and the right kneecap while advancing up Mt Hiei.

The Ashikaga withdrew to Sanjō bōmon, an east–west street located in the north central quadrant of the capital, with the main encampment at the temple of Tōji to the south. They adopted a defensive perimeter. Tōji was heavily guarded and functioned as an imperial palace, for Takauji had the support of a rival to Go-Daigo. The Ashikaga suffered from numerical inferiority, but Takauji had ordered that at all cost Tōji was to be held. Rumours of an immanent attack by Nitta Yoshisada and Nawa Nagatoshi (d. 1336), the two surviving generals of Go-Daigo, led the Ashikaga to set up two forward camps. The first, to the north of the capital, was at Uchino, and commanded by the Hosokawa while the other, in the east and south of Mt Hiei at Hōjōji, was under the command of Kō no Moronao (d.1351).

More than rumours alerted the Ashikaga to the enemy's intentions, for a reconnoitring force launched probing attacks on evening of the 27th day of the sixth month, and watchtowers (*yagura*) located at Sanjō bōmon were ignited, although Kamishiro Kaneharu, a *gokenin* supporter of the Ashikaga, extinguished the flames. It is impossible to know how high these watchtowers were, but even something of three storeys in height would have been sufficient to offer a good view of the flow of battle. Judging from the Ashikaga's ability to react to enemy movements, it seems likely that their watchtower remained, still usable in battle some three days later.

The main force of the enemy, commanded by Nitta Yoshisada (1301–38) and Nawa Nagatoshi, came down Mt Hiei at dawn and circled around to the west, where they soundly defeated the smaller Hosokawa forces at Uchino, and advanced along two parallel streets, Ōmiya, with Yoshisada in command, and Inokuma located just to the east with Nagatoshi in command. There, crucially, a reserve force fighting for the Ashikaga under the command of Shōni Yorinao, chose to remain where they stood rather than fleeing back to Tōji, or directly reinforcing Moronao. This proved fortuitous, for they were able to threaten Yoshisada and Nagatoshi's forces with the unwelcome prospect of being surrounded.

Unaware that they were in the process of being cut off, Yoshisada and Nagatoshi advanced to the east gate of Tōji, where Yoshisada apparently unsuccessfully challenged Takauji to a duel. While this eastern gate remained shut, the Uesugi launched a ferocious counterattack out of the northern gate of Tōji, and in the narrow confines of the capital the Nawa and Nitta forces fled from where they came to the north. The rapidity of the advance meant that few structures had been burned, making the charging horsemen vulnerable to archers stationed in houses or shooting from rooftops. The Shōni forces pursued the retreating army along narrow streets with the Nawa, who travelled along the more easterly road, bearing the brunt of casualties.

RETREAT AND DEFEAT

At the intersection of the Inokuma and Sanjō streets, close to where the main camp of Shōni reinforcements had been

THE BATTLE FOR THE CAPITAL, 1336

The battles of 30 June 1336, revealing the mobility of mounted horsemen in battle. The Southern Court forces of Nitta Yoshisada and Nawa Nagatoshi, depicted here in white, swept down from Mt. Hiei in the northeast. At the onset of battle, they occupied the northern reaches of the capital, and quickly outflanked the Ashikaga troops to their immediate south at Uchino before veering to the west and then south in an attempt to take out their enemy's command center at Tōji. The Ashikaga troops in the north held their ground, and those in the south counterattacked, killing Nawa Nagatoshi and forcing Nitta Yoshisada to flee north. Taking advantage of their victory, the armies of Kō no Moronao, which were encamped on the bank of the Kamo river, advanced to the northeast to the top of Mt. Hiei. Southern forces continued to occupy Amidamine, in the southeast, and their presence ultimately forced the Ashikaga armies to withdraw from Mt. Hiei.

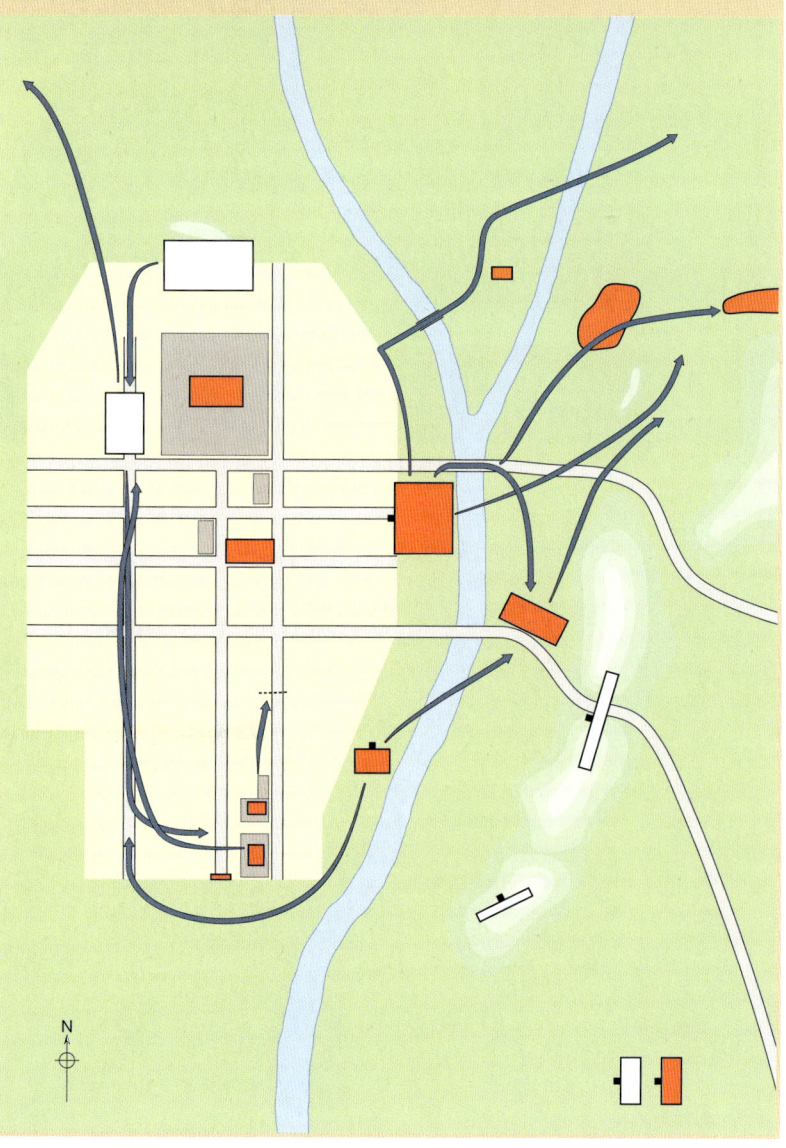

located, Nawa Nagatoshi was killed. The retreat was quite rapid, and more Nawa followers were killed or captured at Oshikōji and Inokuma. One Ogasawara Ujihira wrote how he captured one head, and took one prisoner at Uchino, the site where the Nawa army ceased to exist as an effective force. Yoshisada, located further to the west, was driven back as well, but as the Nawa bore the brunt of the attack his horsemen were able to retreat and flee the capital, along the Nagasawa road to Tanba in the northwest.

In the meantime, seeing the defeat of the main army to the west, scattered warriors joined Kō no Moronao's army, which advanced from Hōjōji to Tadazu near the Lower Kamo shrine. Other warriors expelled enemies from nearby hills, such as Kaguraoka, and then the Ashikaga pushed their advantage and ascended Mt Hiei, taking the fight to the enemy. In spite of killing and decapitating some priests, the Ashikaga were unable to burn the temple of Enryakuji, located on Mt Hiei, and rumours of another attack to the south, coupled with superstitious fears of Enryakuji's power, caused these forces to withdraw. A final attack by the southern forces moved up near Tōfukuji along the road from Toba, and reached as far north as Rokuhara, where fatalities are recorded, but these forces were again defeated and forced to withdraw initially to Daigoji, before fleeing further south. Avoiding overextending their forces, the Ashikaga ended their occupation of Mt Hiei, although they maintained control of the Daigoji region several kilometres to the south of the capital.

The battle proved significant, for afterwards warriors flocked to Takauji to demand rewards, while Tōji temple also received unparalleled grants for its perceived mobilization of divine power. The Southern Court managed to reoccupy Mt Hiei and the hills above Kyoto, but their position remained tenuous. Southern Court partisans continued to occupy Amida peak, located to the south-east of the capital,

THE SKIRMISHERS

A woodblock print of Nitta Yoshisada (1301–1338), the general who sacked Kamakura in 1333. Here, he is offering his sword to the gods so that the tides will lower and he can attack along the beach. He succeeded in this campaign, but was later ignominiously killed by skirmishers when his horse stumbled into a rice paddy.

until late in the eighth month of 1336, when they were defeated and forced to withdraw. As early as the 30th day of June, rumours of their retreat were rife, as indicted by Ashikaga Takauji's letter to Ogasawara Sadamune five days after battle (see panel right). Some literary accounts describe battle formations in the tens of thousands (highly unlikely), while the claim of 1000 enemy killed or captured is plausible. One also finds in a letter from Daigoji reference to Togashi Taka'ie capturing 'several tens of live prisoners and enemies.' As warriors from eastern, western and central Japan participated in this conflict, it seems likely that a force of several thousand warriors was involved in the fray.

The number of documented warriors participating in this conflict, however, proves more limited, for we can reconstruct the actions of a total of 45

> ‘ The Nitta rebels won every battle, but on the last day of the sixth month, we either captured alive or killed several thousand enemy, including [the general Nawa] Nagatoshi, while the forces of Mt Hiei had been diminished. This morning many are fleeing, or surrendering. I heard a rumour that Yoshisada will flee to the east, so thus forces approaching from the east should remain in Ōmi province, and intercept enemy boats and provisions, as well as mop up the fleeing enemy. Post this throughout the roads to the capital. ’

*Tōji temple, located in the southern part of the capital, was where the Ashikaga encamped with their emperors in 1336. During the battles of 30 June, Nitta Yoshisada, according to lore, challenged Ashikakga Takauji to a duel by this eastern gate, but Takauji ignored him. Takauji's supporters instead counterattacked from Tōji's northern gate, and soundly defeated Yoshisada's forces. To this date, the gate pictured is known as the 'unopened gate' (*fukaimon*).*

men, with five Go-Daigo supporters captured alive and four dead for the defeated side. The triumphant Ashikaga had four documented casualties, with two of the wounded dying from their injuries. In addition, there are five cases of *buntori*, which refers to the capture of some part of the enemy or his armour. The aggregate statistics for this battle, excluding vague terms such as *buntori*, reveal that 11 per cent of all warriors mentioned in this encounter were taken prisoner, and 13 per cent were killed, while only 4 per cent were wounded. The percentage of prisoners and fatalities is indicative of a decisive battle, and one that entailed far greater casualties than was typical for a samurai skirmish, but mortality rates remained relatively low.

It is also possible to reconstruct the actions of 18 men who were affiliated to the victorious Ashikaga side and who fought against the Nawa. One of these 18 Ashikaga supporters was wounded, but no other casualties were suffered. These men were able to capture three enemy prisoners, take two enemy heads and kill one further enemy. By contrast, no records survive from warriors fighting for Nawa Nagatoshi or, for that matter, Nitta Yoshisada, indicative of the magnitude of their defeat.

The Southern Court forces managed to continue occupying the Amida mountains to the south-east of the capital, and records survive here of troops affiliated with both armies. Of the 15 Ashikaga supporters who fought in the south, three were killed. They seem to have achieved less success than their compatriots to the north, for we can only document their taking of one prisoner.

Most of the fighting was done on horse, and with bows and arrows. Some warriors competed to be the first to charge against the enemy, while others covered extreme distances in a single day of fighting. Kamishiro Kaneharu, for example, fought at Hōjōji then battled his way to the summit of Mt Hiei, which he spent the evening guarding. Toyoshima Iehide, who fought at Imakumano to the east of Hōjōji, advanced as far south as Daigoji, well to the southeast of the capital, where 'fighting several tens of thousands of mortal enemies [*onteki*] he fought repeatedly, exhausting all of his arrows, and then resorting to blows with his swords, where a number of enemy heads were taken.' Tahara Naosada, of Kyushu's Buzen Province, and therefore a warrior attached to the Shōni forces, describes the nature of the conflict well. While at the Yoshida riverbed, to the east of Hōjōji, his son 'led the charge and dropped an enemy' and continued his pursuit, while a retainer followed and 'captured the head' of this unfortunate opponent, a laconic way of stating that the downed enemy had his head chopped off.

THE LAST DAY

The battle of the last day of the sixth month in 1336 represents a decisive encounter, fought on terrain that was less than ideal for the mounted samurai. Strategically, the Ashikaga gained control of the capital, and their ultimate victory began to be seen as being inevitable, which caused many wavering warriors to rejoin the Ashikaga cause. Some units, particularly the Nawa forces, suffered great casualties and the loss of their commanders, as they were pursued through the narrow streets of Inokuma. Nitta Yoshisada's survival, and the continued entrenchment of Go-Daigo's armies on Amidamine, meant that the southern armies could continue to menace the capital for two more months, until a larger offensive forced them to withdraw. They nevertheless did so in good order, and managed to retain coherence. A relatively high number of prisoners were taken, and fatalities were high, although not as high as cases when whole armies were surrounded and annihilated, such as the battle of Aonogahara in 1338, where Kitabatake Akiie (1318–38) was killed with 700 of his men. Likewise, Go-Daigo's forces fled to the mountains in the south, and managed to weather the death of Nitta Yoshisada in 1338, and cling to power

One of Ashikaga Takauji's most skilled generals, Kō no Moronao, later was unfairly made into a Japanese villain. A woodblock print based on an eighteenth century play, the Treasury of Loyal Retainers, *here shows him anachronistically.*

until 1392, when they finally surrendered to the Ashikaga. But already by 30 June 1336 Ashikaga hegemony was assured. Tactically, this battle witnessed no significant innovations. Instead, it shows that horsemen proved durable riders, capable of travelling many kilometres in battle. They suffered at times in street fighting, but generally proved mobile enough to avoid significant casualties. The most successful advances were along the riverbeds, something that the army of Kō no Moronao guarded against, and took advantage of in their counterattack. Battle, too, was a matter of outflanking enemy forces, and cases where they were overextended, and their horses exhausted, allowed for major counterattacks. At times units could be coordinated, and dispatched to tactically important areas, for the Shōni reserves were able to view the flow of battle from watchtowers. Casualties were high but not ruinous. Warriors fought prudently, and they fought to win. They were not merely staging theatrical appearances, or trying to impress women, as some historians, over relying on anthropological accounts, have suggested.

The fourteenth century witnessed improvements in armour, but no major tactical changes. Battles were of a greater magnitude in 1336, but not fought fundamentally differently from skirmishes of the twelfth century. Sixty years of prolonged warfare necessitated logistical and organizational improvements over the course of the fourteenth century, which in turn would allow for armies to become more cohesive. Armies that trained together proved capable of tactical innovations, and 131 years after the Ashikaga triumphed in the capital, a new style of warring, relying on lightly armed pikemen, predominated, and cavalry no longer dominated the battlefield. In 1467, the mobile fighting of horsemen would give way to defensive tactics. Hence we now turn to a different figure, the pikemen, and explore this new pattern of warfare.

Mount Hiei, the site of Enryakuji, viewed from the Kamo river. Most mounted samurai, when fighting in 1336, advanced along the broad shallow riverbeds in the capital, which provided ample room for horses to run. The narrow streets of Kyoto favoured skirmishers, who could fire down from rooftops at enemy horsemen.

The Pikemen

Pikemen came to constitute the backbone of Japanese armies from the mid fifteenth century onward, but their dominance took time to develop. Major innovations in tactics do not just arise with the appearance of a new weapon, whose use miraculously transforms the battlefield. Tradition and past practices stand in the way of transforming fighting practices, until, often decades if not years after the initial changes, commanders and troops are able to fight in new ways.

Pikemen were the mainstay of Japanese armies, particularly after 1467. Their importance is illustrated in this movie scene from Akira Kurosawa's masterful interpretation of King Lear, *entitled* Ran.

To take one example, the French Army possessed a precursor to a machine gun, called a *mitrailleuse* but could not effectively use it against the invading Prussians in 1870–71, and suffered a crushing defeat. Similarly, pikes had been in use for centuries before commanders were able to realize their full potential. Improved logistical and organizational abilities allowed for standing armies to be forged over the course of the fifteenth century. Once this had happened, troops began to train together, and this led to their utilizing pikes in formation. The next three chapters will discuss the ramifications of these changes, focusing on the use of pikes, improvements in command and the adoption of the gun.

The pike, in many ways, reflected the most important tactical transformation, for it allowed for the rise of what can be called 'massed tactics', whereby men standing in close quarters used their weapons as a coherent group. Such behaviour was not in evidence in the scattered battles of the fourteenth century, where a squad of cavalry remained the dominant battle formation. Too often, historians have thought that guns caused the rise of these massed tactics, but in fact they predominated well before the gun was widely used.

Spears and pikes existed for centuries in Japan, having been used in the Asian continent, and particularly amongst the massed armies of China, but they fell from favour. We only begin to see their return in the fourteenth century. One type of pike, called the Kikuchi pike, is attributed to a commander fighting in 1333, and it consists of short blades inserted in bamboo poles. In fact, similar weapons had been used earlier in the thirteenth century, consisting of knives inserted in poles 1.52m (5ft) in length, for an early image of them appears in the *Scrolls of the Mongol Invasions*. Warriors unable to own more effective or expensive armament seem to have relied upon these weapons. They did not, however, cause considerable casualties, for only 15 examples can be documented over the course of the fourteenth century. If anything, these early pikes were poor men's versions of the curved glaive (*naginata*) and were not remarkably effective weapons.

Pikes became more formidable when numerous men wielded them in close quarters, thereby creating a bristling wall that could not be broken by a squad of horsemen. To create such a force, a commander needed enough cash and supplies to maintain an army in the field, thereby allowing his men to train together so as to march in formation, and coordinate their use of pikes. This necessitated a steady and consistent stream of supplies. Although increased organizational and logistical prowess potentially allowed for units of pikemen to stand up and defeat cavalry forces. Men wearing minimal armour, wielding pikes, could only realize their full potential when some astute commander realized that they could defeat cavalry units on an open battlefield.

Warriors and followers, fifteenth century. High-ranking warriors continued to fight on horseback, but they commanded armies of increasing size, suggested by the large number of men on foot surrounding them.

ASHIGARU (SIXTEENTH CENTURY)

As pikes became the mainstays of Japanese armies, they increased in length, until examples of 5.5m (18ft) in length became common. The warriors here wear *tōsei gusoku* armour of the sixteenth century and are depicted wielding a pike of approximately such length. Oda Nobunaga increased the length of his pikes to 8.2m (27ft), which gave his infantry a profound advantage on the battlefield. The silhouette (bottom left) puts the length of his pikes into perspective – these weapons were unwieldy, but very effective against cavalry.

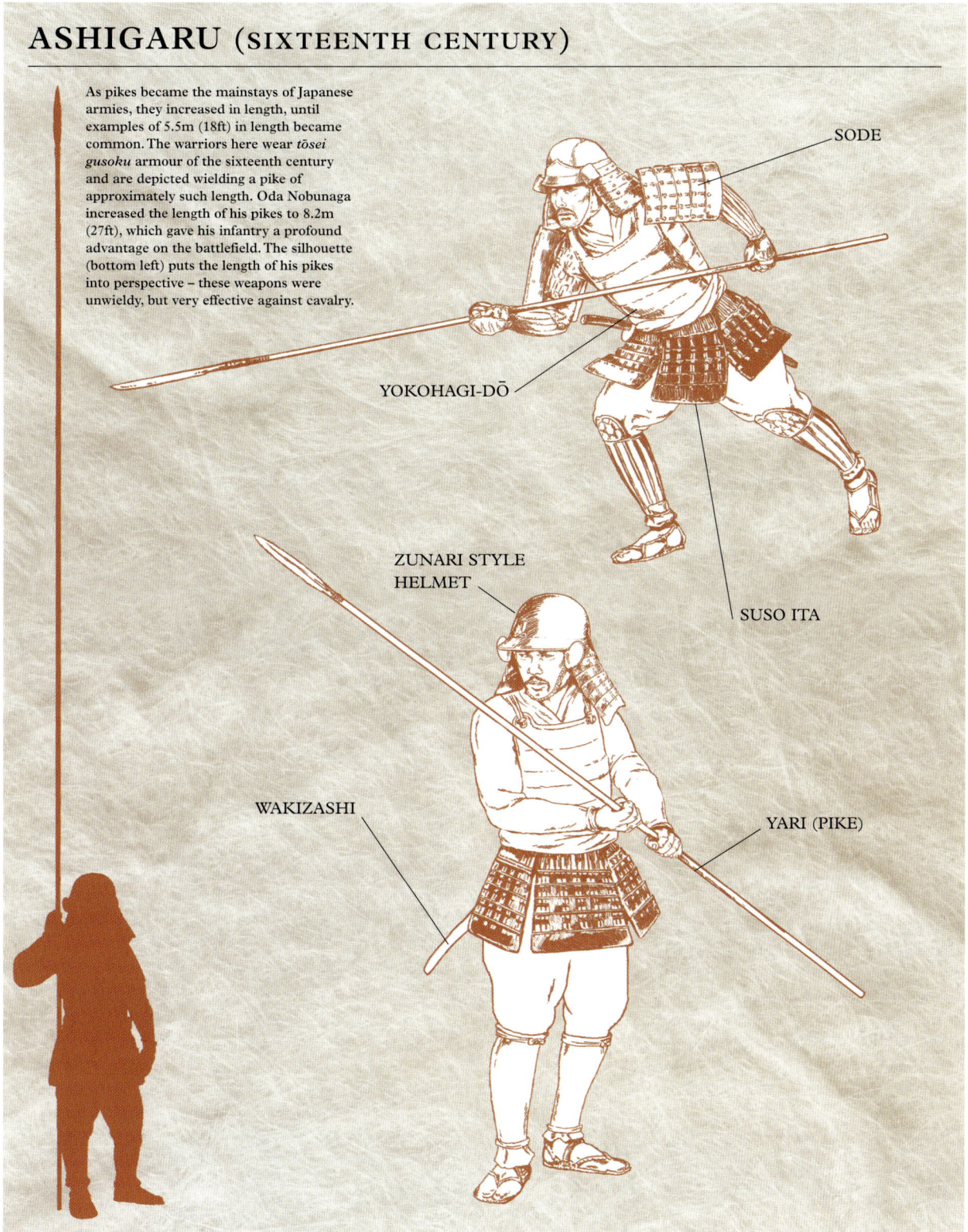

This nineteenth-century illustration reveals how men continued to train and fight with wooden staffs, ably revealing that the longer one's stick, the greater an advantage one had in a mêlée.

During the fifteenth century, a marked and surprising shift in the nature of hand-to-hand combat becomes evident in Japan. Contrary to common assumptions, swords, the so-called souls of the samurai, were rarely used after 1467. While swords generated 92 per cent of all wounds stemming from close-quarters combat (as opposed to all ranges) in the fourteenth century, they were responsible for only 20 per cent of these wounds from 1467 onward. Pikes, which had inflicted 2 per cent of all such wounds in the fourteenth century, caused 80 per cent of them from 1467 until 1600. This preference intensified over time, for pikes caused 74 per cent of all non-projectile wounds from 1467–77 (14 out of 19), and 98 per cent of all such wounds (75 of 76) by 1600. Nevertheless, even though pikes were used so commonly, commanders when describing an encounter of troops with pikes would still be described as an 'exchange of sword blows' (*tachi uchi*), thereby revealing that swords retained a cultural or linguistic significance, even if they were rarely used on the battlefield.

IMPROVEMENTS IN MILITARY ORGANIZATION

By 1467, pikemen had become the dominant figure of the fifteenth-century army. Their significance highlights the greater cohesion of fighting units and improved military organization, as units were capable of training together and remaining together in the field. The establishment of what can be considered a standing army led to profound changes in patterns of warfare, favouring formations of men on foot wielding pikes over their mounted adversaries.

Early fifteenth-century petitions, drawn from eastern Japan, reveal that warriors were increasingly drawn from a single region, and fought together as a cohesive group. During 1417–18, warriors from Musashi Province organized into either the 'Northern White Flag Corps' (*ikki*) or the 'Southern Corps' – geographic origins mattered more than kinship ties. Likewise, the name of one of these troops, the 'White Flag Corps', suggests that its members wore common emblems, such as pieces of white cloth with their names on or common symbols. The oldest surviving example of this practice can be found at the Ōyama zumi shrine. Incidentally, the white flag had no connotations of surrender, as is the case in Europe; instead it revealed a nominal link to the martial Minamoto lineage, the progenitor of most of Japan's most prominent warrior families. By the mid fifteenth century, units appear to have

BELOW, TOP: A jumonji *pike, named because of its cross-like appearance (or the character for the number ten), has two protruding blades, which were designed to block an enemy weapon, or slash to the side. This type of pike became very popular in the Tokugawa era, but was not used often in combat.*

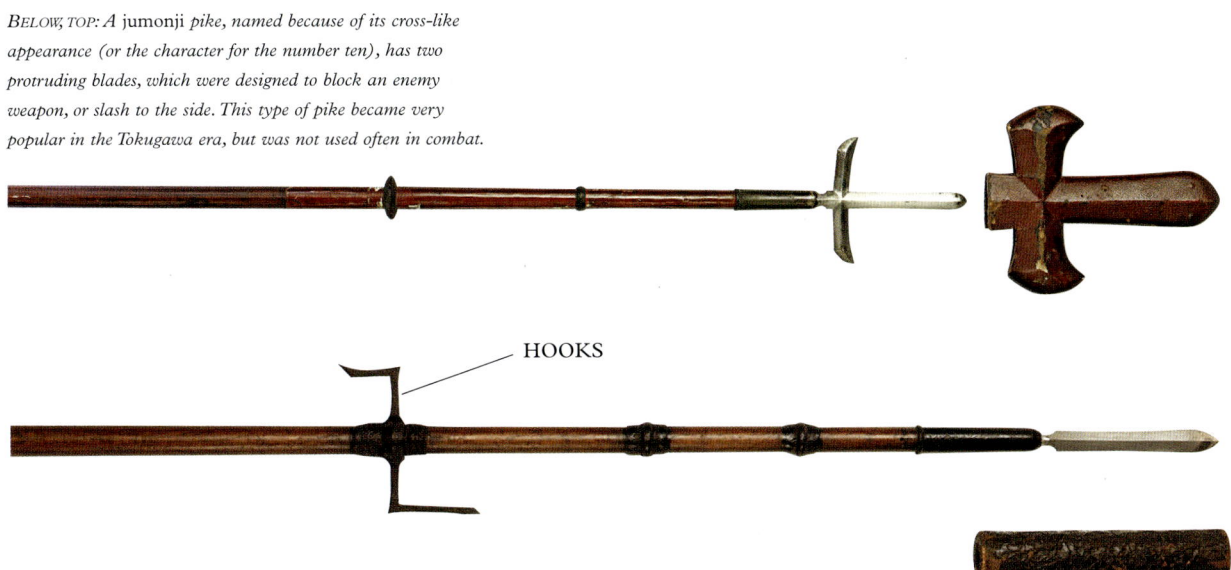

HOOKS

been more organized by geographic basis than appearance. Warriors from the provinces Musashi, Kōzuke and Shinano provinces in central Japan fought as cohesive forces in 1423, and by 1440 generals commonly commanded troops drawn from a single region. This practice is a far cry from the battles of 1336, for example, when troops from a variety of provinces combined to fight together under several commanders.

Surviving records suggest that no tactical transformations arose prior to 1450. Battles continued to be waged as they had been in the fourteenth century: men shot their opponents with arrows, bludgeoned them with swords or hacked their way into fortifications. Tellingly, swords caused all recorded examples of fifteenth-century horse wounds, a trend consistent with the latter decades of the fourteenth century. That horses continued to be slashed reveals that horsemen continued to charge through infantry formations, just as they had in the fourteenth century.

The only visual source of this period, the *Yūki kassen emaki*, shows only minor changes in warrior armour, with more pronounced leg protection, but otherwise armour was virtually indistinguishable from that which was worn before. The warriors depicted, although defending a temple, rely on bows, swords and horses much as they had in the thirteenth and fourteenth centuries. Nevertheless, another surviving scroll, dating from 1433 and from the personal collection of the sixth shogun, Ashikaga Yoshinori, called the *Jingū kōgō engi emaki*, reveals no important changes in armour or techniques of fighting.

By the mid fifteenth century, evidence of improved military organization led to major changes in the practice of war. As armies became more coherent, no need existed for troops to submit documents reporting their arrival in camp. Likewise, when submitting

*ABOVE, LOWER: A 'key pike' (*kagi yari*) became the preferred weapon of the sixteenth century, according to some sources, making up 80–90 per cent of the pikes used in the Osaka campaigns of 1615. In addition to having a stabbing blade, one to two 'keys' or hooks were added to capture enemy pikes. As these hooks were located closer to the hands, they could be more effectively wielded than scythe pikes (*kama yari*).*

POLE ARMS

Most pikes consisted of a relatively simple blade attached to a long wooden or bamboo pole. More elaborate variations contained hooks, designed for catching an enemy blade, or also useful for poking laterally, thereby giving a pikemen more chances to wound an opponent. Scythe-bladed pikes (*kama yari*) allowed for slashing and stabbing while cross-shaped weapons (*jumonji yari*) became most popular in the peaceful Tokugawa era (1600–1868).

KAMA YARI   JUMONJI

WAR BOATS

The *Jingū kōgō engi emaki* does suggest that over the course of the fifteenth century the Japanese were apparently capable of manufacturing war boats, large craft with outlying platforms from which archers could shoot arrows at opposing ships. These sailed craft represent a marked improvement over the craft used in the thirteenth century against the Mongols. According to the *Scrolls of the Mongol Invasions of Japan*, the Japanese defenders could only rely on primitive small boats that were not particularly seaworthy. The later craft proved much larger, and had towers built at the fore and aft of the ship for archers to be placed, as well as areas along the side for archers to stand. These were not particularly manoeuvrable craft. These boats served merely as platforms for many archers, and allowed troops to land on islands or other areas. The existence of such boats helps explain how the Ōuchi, perhaps the most powerful daimyō of western Japan, was so readily able to reinforce their allies in the capital in 1467. Other documents reveal that supplies for armies were sent by boat as well. Documents from the Endō Ishikawa collection likewise reveal how warriors made no clear distinction between fighting on land or on boats. In this sense, boats and horses were largely conceived as mobile platforms from which archers could fire.

In this scene, one can see the difference in size of a small Japanese boat (lower right) with that of the more impressive Mongol ships from China. The man in the boat is holding one of the oldest depictions of what appears to be a pike, used here to help latch onto an enemy boat, although because of the image's poor state of preservation, this object could be a grappling hook, or 'bear claw' (kumade).

petitions for rewards, warriors did not need to list where they had travelled, as the movement of armies became well known to commanders. Changes in military organization led to earlier reports recounting warriors' movements being replaced by simple rosters of wounded soldiers. The need to account for where a warrior had fought became unnecessary once large armies became the norm, as a commander now knew his troops' location. Epitomizing this difference, surviving fourteenth-century documents describe only 721 cases of battle wounds for a total of 8634 individuals. Later lists prove more rare, for only 94 documents survive, but they describe 1208 wounds, which is 67 per cent more than all found in the earlier records. Armies were becoming more coherently grouped, and expanded in size.

THE LOGISTICS OF MILITARY FORCES

Ashikaga Takauji, the first shogun, suffered a particularly difficult time in 1352 as a result of a corrosive and violent dispute with his brother Tadayoshi and son Tadafuyu. In order to surmount these difficulties, and attract the support of the most powerful warriors, he established a tax, known as the *hanzei*, or 'half tax' to be levied in eight war-racked provinces. The *hanzei* enabled men who were appointed as 'protectors' or *shugo*, to use half of a province's tax revenues for military provisions. As time passed, these levies became enforced, and many warriors found that not being able to compete with a *shugo* or constable it was preferable to become a deputy, or otherwise serve in their organization. Gradually, service to a lord rather than autonomy in battle came to represent the basis for success.

The 'half tax' some became an expedient method for *shugo* to amass considerable supplies. One mobilized 755 labourers to help him build a castle, while another used his control over a province to recruit smiths and manufacture arms. It took time, however,

for these *shugo* to break the autonomy of the most powerful warriors in their provinces. To take one example, the Hosokawa family served as *shugo* of Tanba province, and applied continual pressure to the region's most powerful warrior family, the Nakazawa. In 1435, they doubled some taxes on them, but Nakazawa compliance proved limited and grudging, and, even as late as 1445, they occupied contested lands in spite of orders for them to desist.

Nevertheless, at some point over the next generation they abandoned their autonomy and instead cooperated with their *shugo*, becoming managers of the Ōyama estate in 1482, but keeping all the taxes of the lands for themselves, actions that could only have been encouraged by the constable leader. Those *shugo* who managed to control tightly a province's land could then fuse its warriors into a cohesive fighting force, capable of being trained and, most importantly, supplied indefinitely. These changes provided the backdrop for a tactical transformation, and the sudden use of pikes to defeat cavalry formations.

The Hatakeyama, one of the three highest ranking *shugo* families, suffered from intense familial rivalries over who in the family would become heir to their many *shugo* posts, as well as the position of deputy shogun. Bribery led to conflicting judgments by the Ashikaga shoguns, which only served to intensify the dispute and encourage each side to believe in the justness of his cause. Ultimately Hatakeyama Yoshinari (1437?–90) fought his cousin Hatakeyama Yasaburō (d. 1459) in 1454–55, over who would be family heir

*This detail from a screen depicting the 1615 fall of Osaka Castle reveals how pikemen constituted the mainstay of armies, and how two competing armies would attack. Pikemen had to be trained to stand close together and fight in a coordinated manner if they were to be effective. Note too, how simple, rather than scythe pikes (*kama yari*) were preferred.*

and the *shugo* of multiple provinces. Initially the Ashikaga supported Yoshinari, while Yasaburō could rely only on his supporters from Kawachi and Kii provinces, located in the mountains to the south of the capital, along with a few nearby warriors, such as the Tsutsui of Yamato. In spite of an almost complete absence of allied support, Yasaburō fought in a battle whereby some warriors were described as 'being killed as a result of an exchange of pikes.' The language describing this battle is significant, for it suggests a tactical innovation, whereby battles were predominately fought in the mountainous areas of Kii Province with pikes.

His resounding victory allowed Yasaburō to be pardoned by the Ashikaga. If Yasaburō is a military genius he unfortunately left few records or accounts, dying in 1459 before the outbreak of generalized warfare. Still, the changes that he ushered in were epochal, for provincial warriors from Kawachi and Kii could now discourage enemy horsemen from coming near by effectively wielding an inexpensive and inherently fragile weapon effectively, as a single unit on the battlefield.

The Hatakeyama dispute festered, and Yoshinari's fortunes suffered when he mistakenly gave a withered tree to the Ashikaga shogun, Yoshimasa, thereby causing the regime to withdraw its support and instead favour Hatakeyama Masanaga (1442?–93), Yasaburō's heir and younger brother. The Ashikaga even issued an edict demanding Yoshinari's

ASHIGARU (1470)

During the ferocious fighting in the capital in 1470, *ashigaru* plundered temples and residences for wood to be used to make shields or watchtowers, as well as other objects useful in battle. Here, an *ashigaru*, wearing simple *dōmaru* armour and carrying a pike typical for the time, has stolen, or scavenged from the battlefield a slightly damaged helmet, in the older style. Men with no particularly strong links to earlier samurai families had the chance to join the large armies mobilized by *shugo* commanders, and those exhibiting particular skill in battle had opportunities to rise in status.

chastisement in 1460. Enduring seven attacks by the Ashikaga *bakufu* in 1462, Yoshinari now fled to Kii Province, located to the south of the capital, in 1463. Hatakeyama Masanaga returned to the capital, unable to defeat Yoshinari, who had lost the sanction of authority, fleeing to the mountainous regions of Kii province. There he fought Masanaga repeatedly, but neither side was able to achieve victory. Hatakeyama forces, under the command of both Masanaga and Yoshinari, proved to be the harbingers of a new style of fighting. The Hatakeyama dispute also provided the spark for a transformative conflagration, the Ōnin War, which lasted for ten years and resulted in the destruction of the capital, and the beginning of a time when the archipelago descended into a period of almost continuous war. In order to understand their contribution, and the nature of the change, it is perhaps best to turn to a narrative of the the Ōnin War.

ASHIGURU PIKEMAN (1550)

Over the course of the sixteenth century, commanders were able to more effectively use pike in battle. Accordingly they increased in length. By the mid sixteenth century, pikemen commonly wielded long pikes of approximately 5m (18ft), although shorter ones were used as well. This man wears a simple helmet, and armour consisting of a cuirass made of metal plates (*yoko-hagi-dō*) and *suso ita* to protect the legs, using minimal braiding. As units became standardized, pikemen tended to wear simple but standardized armour, often emblazoned with the family crest of their *daimyō* lord.

THE ŌNIN WAR 1467–77

Sometimes, greater institutional and organizational abilities leads to greater instability, for as certain offices gain in influence so too do they become more desirable, and thus the focus for conflict. The fact that the position of *shugo* allowed its holder to have access to half

ASHIGARU EQUIPMENT (SIXTEENTH CENTURY)

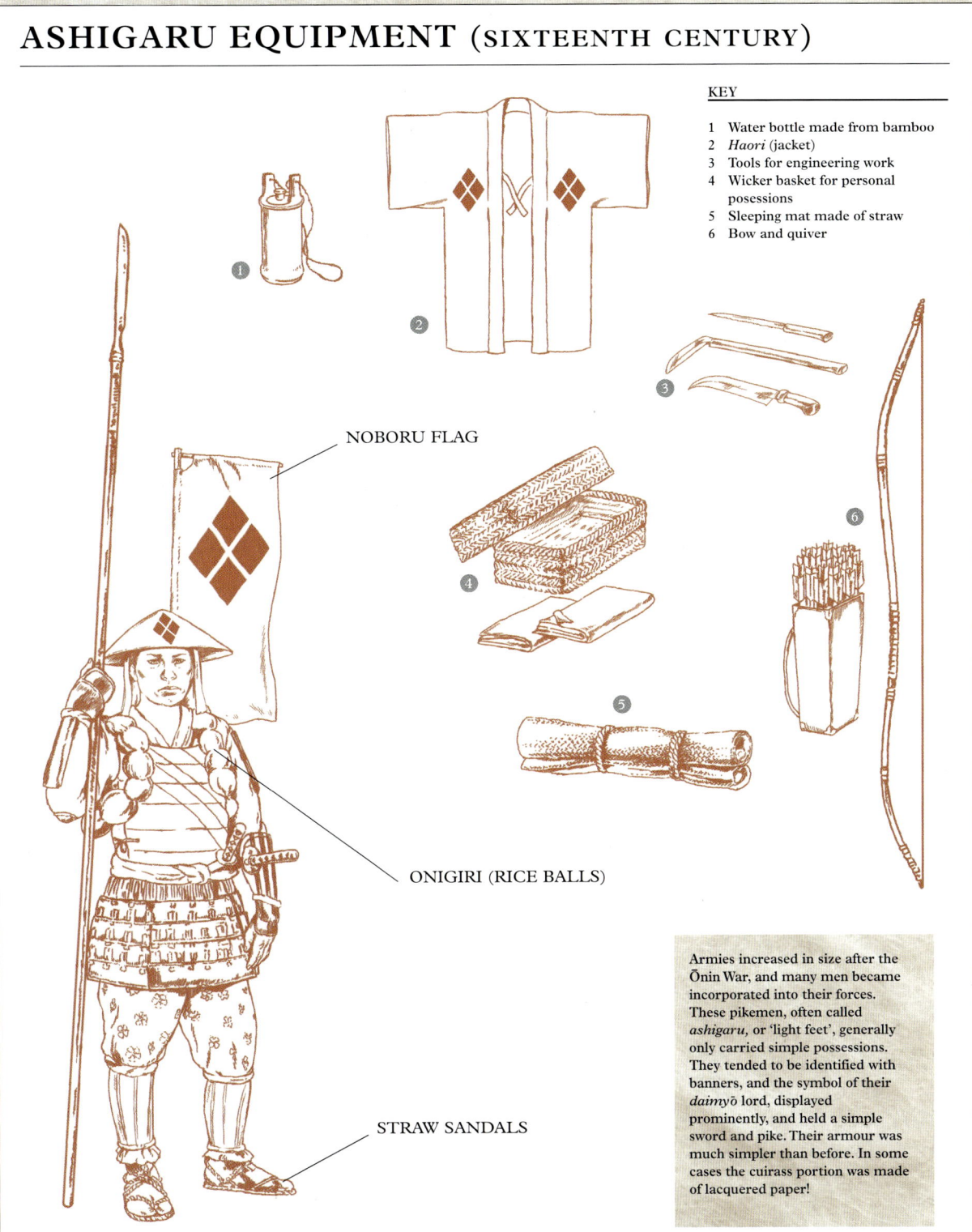

KEY

1. Water bottle made from bamboo
2. *Haori* (jacket)
3. Tools for engineering work
4. Wicker basket for personal posessions
5. Sleeping mat made of straw
6. Bow and quiver

Armies increased in size after the Ōnin War, and many men became incorporated into their forces. These pikemen, often called *ashigaru*, or 'light feet', generally only carried simple possessions. They tended to be identified with banners, and the symbol of their *daimyō* lord, displayed prominently, and held a simple sword and pike. Their armour was much simpler than before. In some cases the cuirass portion was made of lacquered paper!

The shugo daimyō of the fifteenth century proved very effective at procuring taxes, manifested here in rice, and transporting it to the capital. So effective were they in feeding their armies that when war broke out in 1467, these generals could maintain large armies in the capital for a decade. Rivals unsuccessfully strove to block enemy supply lines to the capital to defeat their opponents.

of a province's revenue after 1352 contributed to the desirability of this position; at the same time, it exacerbated rivalries because only one individual could occupy this position.

With the increase of revenue, and the expansion of house organizations and bands of followers known as retainers, factions invariably formed between rival candidates, which could lead to internecine conflicts. The Hatakeyama proved to be a family with the deepest animosities and the greatest tactical skills. Their love of warfare was only matched by an almost congenital inability to compromise. Ultimately the weak-willed eighth Ashikaga shogun, Ashikaga Yoshimasa (1435–90), despaired of ever reaching a settlement over the Hatakeyama inheritance dispute, so he decreed that both Masanaga and Yoshinari were to collect their supporters and fight it out in the woods of the Gōyrō Shrine. They did, and this time Yoshinari triumphed. Reneging on an agreement that no other parties were to interfere, Masanaga's friend, the powerful deputy shogun Hosokawa Katsumoto (1430–73), aided him. This caused others, particularly those friendly with Yoshinari, to protest, and preparations

THE PIKEMEN

began for a campaign to attack the oath-breaker Katsumoto who, because he was the most powerful man in the Ashikaga regime, had a formidable base of support of his own. Each *shugo* had a mansion in the capital, and each fortified his dwellings, relying on his control of the provinces to ensure that a steady stream of men and materials flowed from the provinces to the capital. By a quirk of geography, most, if not all, of Yoshinari's supporters resided in the western wards of the capital, while Katsumoto and his followers were in the east.

Katsumoto, who had been preparing well for a conflict, commenced hostilities on the 26th day of the fifth month of 1467, as he consigned the abode of Isshiki Yoshitada to the flames. Yoshitada's residence was vulnerable, for it was the only mansion of the anti-Hosokawa forces located in the eastern wards of the capital, and the one place where their supporters could contact the Ashikaga shogun. The anti-Hosokawa faction, led by the Yamana, thereupon demolished Hosokawa houses located in the western wards, and the civil war

Until its destruction in the Ōnin War, Kyoto remained the seat of the Court, the Ashikaga shogunate and the centre of most of Japan's wealth. Moneylenders were common here, as too were smiths and armour manufacturers. Something of the vibrancy of Kyoto is evident in this screen depicting scenes from the capital.

TRAINING WITH PIKES

In the mountains of Kii Province, the Hatakeyama first trained groups of men to fight as units of pikemen. Their innovation profoundly influenced the waging of the Ōnin War and led to a tactical stalemate, as squads of mounted archers could no longer control the battlefield. After the inconclusive end of the Ōnin War, commanders strove to train as many men as possible into units of pikemen. This illustration reveals the training of such a units in the mid sixteenth century.

PIKE AND SHIELD

Pikemen were best equipped to physically occupy territory, and defend it with defensive fortifications, such as trenches or earthworks. They also hid behind shields made from wooden boards, as depicted here. Mounted horsemen would not challenge them, although in the fighting of the Ōnin War, rival forces used flaming arrows and projectiles shot from catapults to try to dislodge pike-wielding *ashigaru*.

RIGHT: *Here we see two* ashigaru, *one of them holding a decapitated enemy head. They hold pikes as well as small swords, and their provisions, rice balls (*onigiri*), are draped around their necks.*

had begun. Both sides fought over the capital, for this is where the *shugo* residences had been fortified, and the army that fled the capital feared being classified an enemy of the court, which would entail a loss of legitimacy and an ultimate withering of support. A stalemate quickly arose.

The absence of a decisive victory caused Ashikaga Yoshimasa to treat the conflict as a mere private struggle, and so he largely ignored the fray, preferring instead to engage in more cultured pursuits, ultimately devoting his interest to constructing a remarkable pavilion that is known to posterity as the 'Silver Pavilion' (*Ginkakuji*), one of the cultural glories of Japan.

During the initial two days of conflict, both armies burned much of the capital, for they strove to create wide-open spaces for their horses to roam. All *shugo* dwellings located in indefensible positions were either destroyed or abandoned, with the 'Eastern' Hosokawa and 'Western' Yamana forces suffering the losses of three *shugo* mansions each, along with seven major temples and countless other dwellings. The Hosokawa managed to control the northeastern quadrant of the capital, where the imperial and shogunal palaces were located. Yamana Sōzen (1404–73) responded by calling in supporting troops, who smashed through a depopulated province in the northwest, thereby reinforcing his position in the capital. Three more *shugo* residences and the homes of two nobles were burned on 25 June 1467. After much of the capital had been burned, both armies struggled

THE PIKEMEN

Inland Sea, from his base in western Japan during the eighth month. The newly reinforced Western army burned Sanbōin, a major temple to the southwest of the capital, on 13 September, and cut another of the Hosokawa's supply lines to the capital.

Sources reveal that as early as 13 September pikes were widely used, for six members of the Kikkawa family were stabbed with these weapons, while four more were wounded by pikes on the second and third days of the tenth month of 1467. Not all wounds were caused by pikes, for one Kikkawa family member was slashed with a sword, five were hit by rocks and eight shot by arrows, but the increase is suggestive of a major change, for this one encounter witnessed nearly as many pike wounds as had the previous century of combat. In spite of yet more evidence of a tactical change, the Western commanders still believed that they could break the Eastern army, and occupy the whole of the capital.

The triumphant Western army next began a major offensive with the goal of crushing the resistance of the flagging Eastern army. Having refrained from an offensive campaign in the city, the

PIKE VERSUS SWORD

A pike strike. The greatest advantage of the pikes was that it could kill opponents at a greater distance than other shock weapons. A sword-wielding samurai, when confronted with a squad of pikemen, was defenseless and doomed. Here, a samurai has been pushed to the ground by pikemen who are about to complete the kill.

to control supply lines. During a major campaign in the south-west of the capital, and the Western forces proved successful in defeating the Eastern army in the hills around the temple of Nanzenji, thereby further constraining the Eastern army. The Western army was bolstered by Ōuchi Masahiro (1446–95) leading a large force of warriors via the

Western commanders believed that their cavalry could take the day, and so the area around Shōkokuji was burned on 3 October, so as to provide cavalry with room to roam, but they were decisively defeated by those of Hatakeyama Masanaga's forces who had sided with the Hosokawa.

According to the *Chronicle of Ōnin*, Hatakeyama Masanaga's army suddenly and dramatically defeated a force of Rokkaku cavalry on the burned-out grounds of Shōkokuji Temple, located in the centre of the capital. Masanaga purportedly had a force of 2000 troops, which attacked a significantly larger cavalry force of 6000 to 7000 troops. Masanaga was confident of his prowess for he boasted 'I will defeat an army of even a million men' and his troops advanced in close formation behind shields. After closing in on the Rokkaku horsemen on the ashy grounds, Masanaga's pikemen surged towards the

A six-fold screen depicting the battle of Ichi-no-tani in 1184, dating from the seventeenth century. This anachronistically portrays a small settlement as a town, somewhat evocative of Kyoto. The armour styles are more typical of later ages than the twelfth century.

enemy, whose horses could not withstand such a unit of pikemen – 67 horsemen were killed before the rest fled. Masanaga could not, however, pursue these forces, because another group of pikemen, commanded by the redoubtable Hatakeyama Yoshinari, approached the burned-out temple grounds and forced Masanaga to withdraw. Thereupon the remains of Shōkokuji were deserted.

The dominance of a force of pikemen meant that defensive tactics replaced offensive ones. Mobility ceased to matter as much as the ability to physically occupy contested grounds. The Eastern army began constructing impressive trenches early in 1468, and the Western army soon followed suit. Some trenches were 3m (9ft 10in) deep and 6m (19ft 8in) wide, thus causing Japan's capital to resemble the Western Front of World War I. Watch towers ranging from 21 to 30m (69–99ft) in height watched over enemy positions and became the focal point for attacks and raids. Each army struggled to call up men from the provinces in increasing numbers to man these positions, and conduct raids on enemy outposts. Armies would resort to rocks and flaming arrows to burn down watchtower structures, although they met with limited success.

MINOR ACTIONS

Troops huddled in trenches, with little ability for offensive action, and many of

the battles appear to have generated intense boredom, causing some warriors to engage in poetry competitions. Night raids came to be favoured, as small groups of foot soldiers strove to infiltrate enemy lines. Mobile squads of *ashigaru* burned vulnerable enemy fortifications, or demolished dwellings housing enemy troops. The increase in army size meant that manufacturing could not keep pace, and most of these skirmishers carried shields made of wood and bows made of bamboo, which could be constructed relatively easily. Some accounts describe these men as wearing helmets, while others stated that they 'had no pikes, nor armour and carried only swords.' It seems likely, however, that *ashigaru* functioned more as skirmishers, or irregular fighters, than organized formations of infantrymen.

Even though irregular *ashigaru* were used in the conflict, both armies required immense amounts of armour, and weapons. Provincial centres proved inadequate to keep up with demand, and in addition the best armour seems to have been made in the capital. So valued was this manufacturing region, located in the south of the city, that neither army damaged its buildings, meaning that the great temples, palaces and residences of shoguns, nobles and warriors were largely destroyed in the northern reaches the capital, while the humbler abodes of manufacturers to the south remained unscathed.

Calvary, unable to engage in frontal attacks, instead engaged in raids against the villages around the capital in order to disrupt their supply lines, and also specialized in reconnoitring. Throughout 1468, cavalry tried to cut enemy supply lines and targeted sparsely settled areas so as to prevent the movement of men and supplies. The Western army managed to close down all but one road into the capital for the Eastern army, but this one road proved adequate to keep Hosokawa Katsumoto's army in the field.

NEW TECHNOLOGY

The establishment of a tactical stalemate caused for innovations in technology, as a craftsman from Izumi province built a trebuchet, known as a *hō*, capable of launching a 3kg (6.6lb) projectile for over 274m (300 yards). One also sees references to guns, even though not all knew the words to describe them. Sources describe a 'flying projectile fire spear' that was launched from a besieged tower on 6 November 1468 (for more on these firearms, see Chapter 5).

Nevertheless, guns did not have a decisive role in the conflict, nor did they cause the tactical transformations, namely the rise of a massed army composed largely of foot soldiers and the

THE ŌNIN WARS, BATTLE FOR THE CAPITAL, 1467

Offensive and defensive conflicts, 26 May 1467. The onset of the Ōnin War quickly devolved into a stalemate. Here, attacks by both the Eastern and Western armies are depicted in white, while defensive units are shaded. The Eastern army attacked the Western forces from the north and the east, but were rebuffed, while, to the contrary, the Western armies attacked Eastern supporters to their south. The Western defences held against the Eastern army, and as time passed, and Western reinforcements arrived, the Western army was able to seize control of the southern quadrant of the capital.

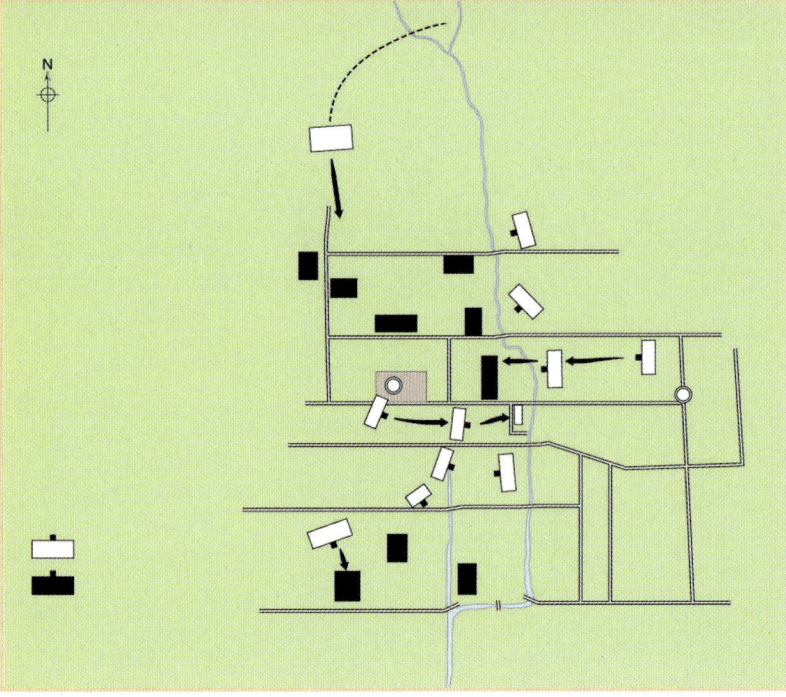

Ashikaga Yoshiteru, the Ashikaga shogun who had the dubious distinction of being killed in an attack on his palace by a daimyō *named Miyoshi Nagayoshi. Nagayoshi symbolizes the spirit of* gekokujō *or 'lower supplanting superior' because he had no qualms about attacking and killing his nominal lords. Such action was difficult to conceive of even during the Ōnin War of 1467–77, as none would dare attack the Ashikaga shogun Yoshimasa. Yoshiteru acquitted himself well in the end, fighting furiously before being overwhelmed. He is depicted here in court robes, and an imperial guard's sword (*efu no tachi*) in this 1868 Yoshitoshi print from the popular series 'One Hundred Warriors'.*

pre-eminence of defensive tactics. Most authors see the arrival of Portuguese firearms as causing a technological revolution, and an expansion in armies and increased use of defensive structures. Instead, the ability to sustain a large army led to a tactical revolution, the use of pikes, which in turn led to increased use of fortifications. The triumph of Miyoshi Nagayoshi (1522–64), a deputy *shugo*, over his Hosokawa lords in 1547 epitomizes the fact that coherent armies, and disdain for the existing social hierarchy, existed independently of the use of Portuguese firearms.

Without ready access to these new weapons, Nagayoshi was able to defeat Hosokawa Harumoto, his lord, with a formidable force of 900 pikemen, and expel the shogun, Ashikaga Yoshiteru (1536–65), from the capital in 1549. Nagayoshi based his position solely on his military prowess, which above all stemmed from his effective use of pikemen. Over the course of the sixteenth century, the need to organize, sustain and command armies proved the basis for military power.

Miyoshi Nagayoshi, in this sense, represents the final legacy of the Ōnin War. For the commanders of the Eastern and Western armies, unable to tactically defeat their opponents in a pitched battle, as well as throttle the flow of supplies to enemy encampments, strove to undermine regional authority by causing deputies in the provinces to rebel against their *shugo* lords. A few deputies, or for that matter, regional organizations of provincial warriors, succeeded in overthrowing their *shugo*, which ultimately caused the surviving combatants to end the hostilities and flee the capital in 1471, and from there to shore up their shaken authority. Those with charisma, and the best ability to mobilize forces and seize the initiative, proved most successful, and so it is to the commanders that we shall turn next.

COMMANDERS

Japan witnessed nearly continuous warfare for centuries. Extended conflicts can be verified as occurring during the years 1221, 1274, 1281, 1331–92, and armed uprisings likewise erupted in 1399, 1413–18, 1422–25, 1428–29, 1431–38, 1440–44, 1451–56, 1459–63 and 1465. After the cataclysmic ten-year Ōnin War, continuous warfare continued thereafter, particularly after a coup in 1493. In spite of the ubiquity of war, few if any of the commanders of these conflicts have been remembered. By contrast, during the latter half of the sixteenth century several samurai commanders became famous: Takeda Shingen (1521–73), Uesugi Kenshin (1530–78) and Oda Nobunaga (1534–82). All of the noted leaders were active during the latter half of the sixteenth century, and their fame and the relative anonymity of their predecessors reveals a changing notion of command, leadership and military control.

A recent re-enactment of Takeda samurai, with commanders seated in front of Takeda banners.

The Ashikaga about to attack Nitta forces during the fifth month of 1336. The Nitta flags are visible in the foreground. The Ashikaga triumphed, and managed to occupy the capital in the sixth month of 1336. This woodblock triptych is by Udagawa Kuniyoshi (1798–1861).

One reason for the anonymity of the early commanders is that many of the conflicts, particularly those during the first half of the fifteenth century, were limited, sporadic affairs. By listing off the years of conflict, one would assume that warfare was constant, but in fact most disputes involved family members fighting over who would have the post of *shugo*, and involved few fatalities. Some of these conflicts were of greater magnitude, but even the Ōnin War did not leave any legendary generals.

Getting men to obey orders, and risk their lives, proved to be the greatest difficulty for commanders. It was not easy to get the free-spirited warriors of the thirteenth century, who valued their lands and autonomy above all, to follow orders. Samurai desired a name and recognition, and so holding back to support an attack proved unpopular. A local commander ordered warriors to wait before charging when the Mongols invaded, but Takezaki Suenaga could not countenance delay. Stating that the way of the warrior is to be worthy of reward, he advanced and was promptly shot from his horse. He still demanded rewards for being the first. The Kamakura *bakufu* issued repeated orders during the Mongol invasions, that warriors were to follow their commanders, but the necessity of such orders reveals the absence of a coherent chain of command. Of course, even generals did not invariably follow their orders. One great difficulty, from a military standpoint, is that warriors who killed or otherwise incapacitated an enemy would strive to cut off his head, so as to provide proof of their valour. This led some, such as Kikuchi Takefusa, to scour the beaches of northern Kyushu after a typhoon struck the Mongol fleet, securing numerous heads of the dead Mongol warriors. Warriors who took a head would rush from the fray in order to show it to a commander and be eligible for great rewards. In 1338, commanders issued a 'cut and toss' order, demanding that warriors refrain from cutting off heads and instead rely on witnesses, but when the enemy general Kitabatake Akiie was killed, the warrior who killed him and another who took his head were rewarded equally.

The generals of the late thirteenth and fourteenth centuries did not command armies as much as collect them. Continually they had to promise rewards to warriors to keep them in the field. Skilled commanders, such as Ashikaga Takauji, were very good and creative at rewarding his followers and ensuring that his promises were kept. By contrast, his wayward son Tadafuyu raised an army by promising all warriors whatever they wished. He quickly amassed a great army, but it soon scattered because Tadafuyu issued contradictory promises, granting as reward the same parcel of land to two different warriors.

Land proved to be a difficult mechanism with which to maintain armies, and so commanders relied on giving other objects. Ashikaga Takauji sometimes granted prized swords to his followers, and at other times, would allow them to use his family crest. Even simple gifts could profoundly influence a warrior. Arai Hakuseki (1657–1725), a Confucian scholar and advisor to Tokugawa shoguns from 1709 to 1716, wrote in his autobiography about how his grandfather, who fought late in the sixteenth century, cherished a pair of chopsticks that his commander gave him after one battle. Still, the greatest gifts that such commanders could give were pieces of paper that confirmed land rights and represented a special link to authority.

Takauji took great care in writing his documents, and he devised a distinctive and artistic signature that became the template for all warrior signatures for the next century and a half. At times, he also relied on a special steel-blue colour of ink that his predecessor, the great Minamoto Yoritomo had purportedly used in the twelfth century. He did not fear death, or so it is said, but he also did not physically lead forces in battle. To turn to the battle of 30 June 1336, Takauji remained the whole time in Tōji,

Edo castle, abode of the Tokugawa shogun, or tai-kun *(tycoon) from 1603 to 1867. The Tokugawa increasingly relied upon advisors such as Arai Hakuseki to govern the realm.*

and when challenged by Nitta Yoshisada he chose to ignore him. This action earned him no ignominy; to the contrary, Tōji boasted that this gate would never open again. Takauji also would compose poetry during battles, and at his most unflappable he commented that the escape of the newly captured enemy emperor, Go-Daigo, late in 1336 was a good thing, because it freed him up from the need to guard the sovereign.

Takauji also commissioned prayers, and drew pictures of Buddhist divinities (*bodhisattvas*), but does not seem to have actively fought much on the battlefield. His ability to reward followers and keep his promises made him a formidable leader. Most remarkably, he once became embroiled in a succession dispute with his younger brother, Tadayoshi (1306–52), and was defeated in battle in 1351, ending up with only 42 followers. He 'surrendered' to his brother, but then demanded that he first be able to reward his men. His brother granted this request, and thereupon Takauji acted as if he had won the conflict, because he rewarded his men first. Such a seemingly semantic distinction proved significant, for it showed that Takauji gave the most valued and enduring gifts. When Takauji thereupon plotted against his brother, and fought again, most warriors sided with him.

Takauji's method of command proved effective, for he established a dynasty that lasted for 237 years. His largesse and gift-giving skills were essential in an age where armies had to be collected, rather than simply commanded. As warriors themselves darted to and fro and were not readily organized, any attempts to micromanage the battlefield would end in failure. Instead, troops were dispatched to particular regions, and they then fought as they saw fit.

FLEXIBLE RULERS

A certain flexibility and willingness to compromise proved necessary for those aspiring to rule the archipelago. The most powerful warriors tended to be the most protective of their interests, while weaker ones tended to side more tenaciously with those in authority. Takauji, the most successful leader of his day, proved adept at balancing the interests of both, promoting weaker allies and distant relatives, and at the same time, enticing the most powerful warriors to join his cause. He was above all a gift giver, who cherished the right to give land rights, titles and names or privileges to his followers. He cared little about direct control of territory, but to the contrary would soon attack anyone who attempted to bestow largesse on others, as his son Tadafuyu discovered to his misfortune. Takauji also had no qualms about attacking his son, who had

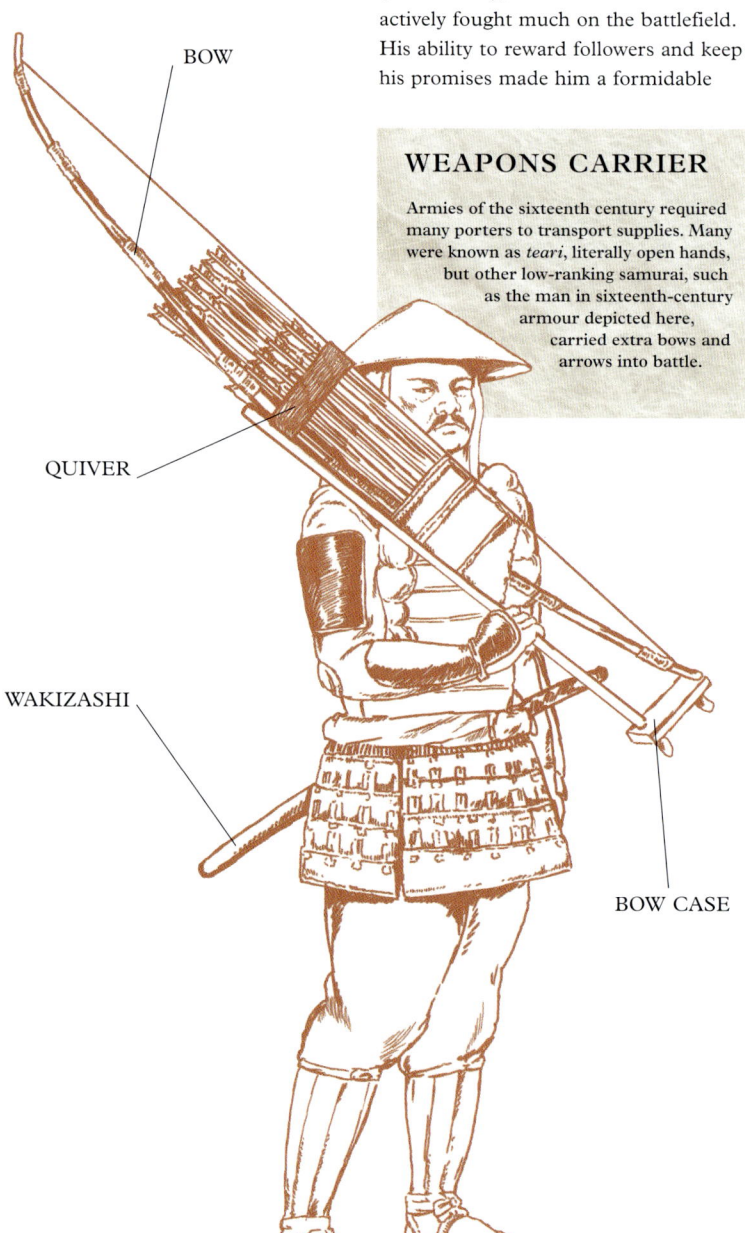

BOW

QUIVER

WAKIZASHI

BOW CASE

WEAPONS CARRIER

Armies of the sixteenth century required many porters to transport supplies. Many were known as *teari*, literally open hands, but other low-ranking samurai, such as the man in sixteenth-century armour depicted here, carried extra bows and arrows into battle.

attempted to usurp his authority by unilaterally confirming lands.

Takauji's greatest gift, known as the *hanzei*, allowed half of the revenue of a particular province to accrue to the *shugo* of a province. This transformed the ability of warrior lords to maintain armies, as we have already seen. Some of the warriors who received the greatest largesse, such as the Yamana, who at one time were *shugo* of 11 of Japan's 66 provinces, were those who fought for both sides, securing the best deal as a result of repeated shifts of allegiance.

Takauji's son, Yoshiakira (1330–67), proved less successful in cobbling together a coalition of warriors. Gradually, the most powerful families joined the Ashikaga over the course of the 1360s, but the stress of keeping them happy, combined with generous drinking, caused Yoshiakira to die of high blood pressure, when an unstoppable nose bleed killed him at the age of 38.

Such a situation proved unstable and required continual cajoling and communication to keep recalcitrant warriors in line, or invested with a particular regime. One innovation is that commanders, who could not always grant land grants immediately, became writing letters of praise. Aso Harutoki (d. 1334) would be the first general to do so, writing on 21 April 1333: 'While skirmishing in the mountains north of Chihaya castle, you took [an enemy] head. This is most splendid.' Harutoki,

SASHIMONO

As armies became larger in size and more coordinated, troops wore streamers on their backs for identification. A member of each company or unit of warriors held a flag in a holder attached to the back of the armour for identification, which was known as *sashimono*. Most late sixteenth-century armour had easily identifiable crests to show allegiances as well.

Detail of a samurai decapitating a prisoner, from a scene from the thirteenth-century Heiji monogatari emaki. *Warriors required proof of service in battle and such grisly trophies were a means of getting ahead in the world.*

SUNBURST STANDARD

The trappings of office and status became more pronounced for sixteenth-century commanders. The sunburst standard represents a later decorative standard favoured by some.

by writing so many letters to his troops, managed to keep his army unified. He faced unusual problems, in that his force had been besieging an enemy castle and while doing so uprisings caused the regime he had been fighting for, the Kamakura *bakufu*, to be annihilated. Harutoki managed to keep his army together, in no small part by winning the respect of his men, and this army remained in being until it negotiated a surrender with Go-Daigo's new regime.

Other commanders maintained their authority through a steady stream of letters. Imagwa Ryōshun (1326–1420), for example, had no ties to Kyushu before he was appointed to command Ashikaga forces in Kyushu in 1370. Dispatched there the following year, he wrote numerous letters or dispatched other documents to local warriors, many of whom had never received such detailed correspondence before. The research of Kawazoe Shōji reveals that 472 of Ryōshun's documents survive, with 132 of these representing detailed letters to warriors under his command. Through these writings, and his exhortations, he managed to secure the support of local warriors for nearly a quarter of a century. Ryōshun also wrote literary critiques and penned poetry manuals, and does not correspond to one's typical image of a warrior leader.

One episode in Ryōshun's career exemplifies the limits of commanders who demanded that their soldiers and allied commanders unthinkingly obey. He invited one duplicitous commander named Shōni Fuyusuke to a banquet and had this man executed for insubordination. This act served, however, to undermine his authority, as many of his allies left and his forces were severely weakened. He did manage to restore his authority over the next 20 years, and he proved largely successful because he refrained from any further executions. Takauji, by contrast, was a more popular commander, because he was famous for not holding a grudge and rewarding even enemies. Clearly, no notion of loyalty existed, or was expected, among the *gokenin* warriors of the fourteenth century.

THE ASHIKAGA PATTERN OF RULERSHIP

Takauji's grandson, Ashikaga Yoshimitsu (1358–1408), proved more adept at securing loyalty, but he did so by his mastery of court ritual. As an infant, he experienced the turmoil of civil war, for he was forced to flee to the provinces at the age of five, and this experience left him with an uncanny instinct for the successful wielding of power. Yoshimitsu, being an infant, had a protector appointed for him, named Hosokawa Yoriyuki (1329–92), who was known as the deputy shogun, or *kanrei*. Yoriyuki strove tirelessly to enhance Ashikaga prestige and authority, and was rewarded in 1379, when Yoshimitsu seized the reins

of power, with banishment from the capital. Thereupon, Yoshimitsu allowed three competing families, the Hatakeyama, Hosokawa and Shiba, to be potentially eligible for the position of deputy shogun, which meant that they constantly competed against each other. Other less prestigious *bakufu* positions became the prerogative of four families, including among them the Yamana and the Akamatsu, and this in turn exacerbated the rivalries. Yoshimitsu thus enshrined a system in which warrior families competed against each other, and at the same time, as all power was invested with the figure that held the office of *shugo*, intra-familial disputes likewise became more common.

Yoshimitsu acted more like a courtier than a shogun, and with his mastery of court ritual he overawed the *shugo*, as well as court priests and nobles. He would travel on lavish journeys throughout the land, so as to scout out the powers of the *shugo*, and at the same time, he would require immense and provisional levies to help him build a variety of palaces and temples. Yoshimitsu also had a palace built called the 'Palace of Flowers', which overawed the imperial palace. He had a temple built called Shōkokuji in 1399, in celebration of his 42nd birthday, which had a seven-storey pagoda and thereby dwarfed all other structures in the capital. Finally, he constructed the Kinkakuji (Golden Pavilion) that survived until a crazed monk burned it down in the 1950. Kinkakuji epitomized the grandeur of

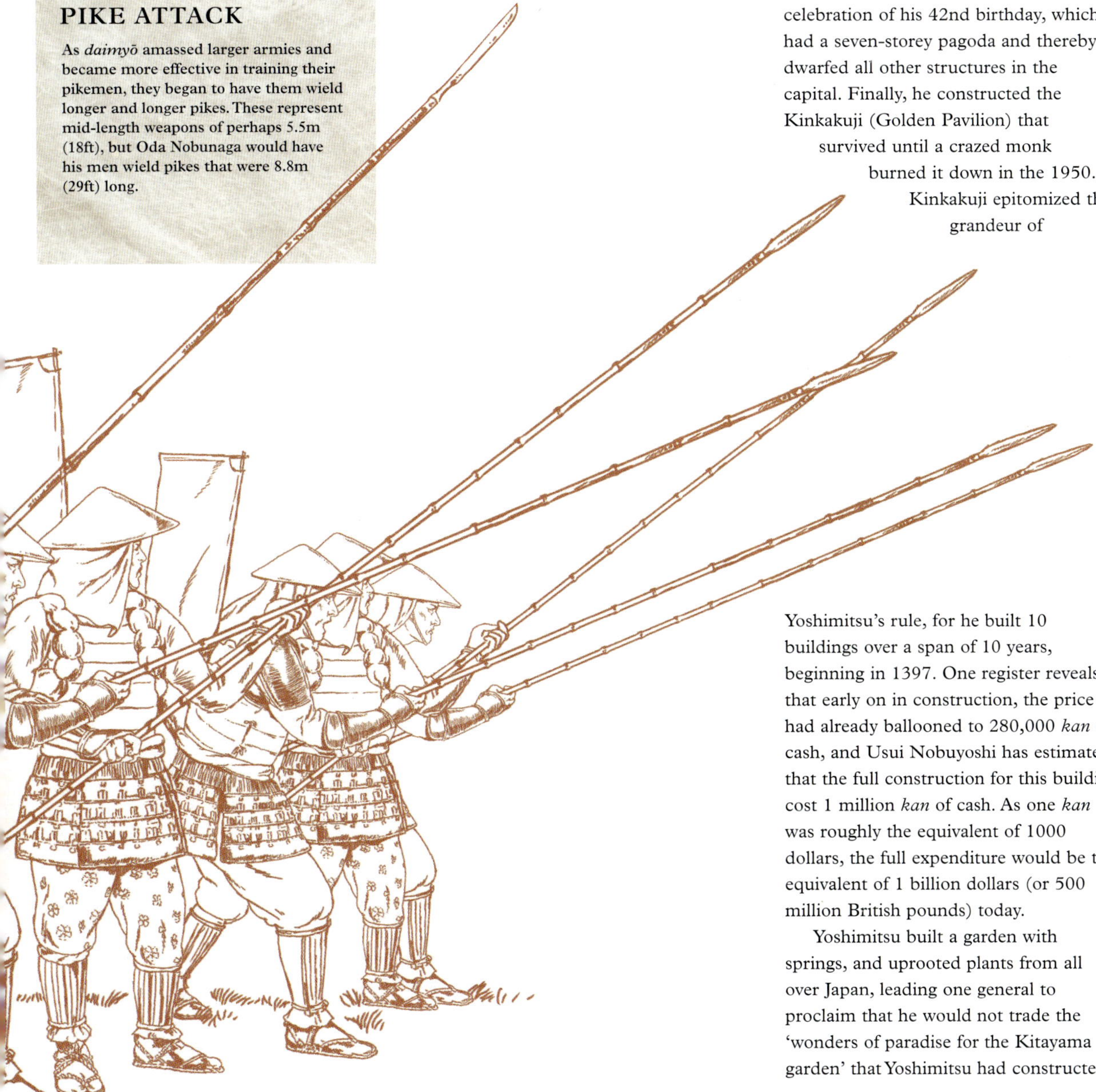

PIKE ATTACK

As *daimyō* amassed larger armies and became more effective in training their pikemen, they began to have them wield longer and longer pikes. These represent mid-length weapons of perhaps 5.5m (18ft), but Oda Nobunaga would have his men wield pikes that were 8.8m (29ft) long.

Yoshimitsu's rule, for he built 10 buildings over a span of 10 years, beginning in 1397. One register reveals that early on in construction, the price had already ballooned to 280,000 *kan* of cash, and Usui Nobuyoshi has estimated that the full construction for this building cost 1 million *kan* of cash. As one *kan* was roughly the equivalent of 1000 dollars, the full expenditure would be the equivalent of 1 billion dollars (or 500 million British pounds) today.

Yoshimitsu built a garden with springs, and uprooted plants from all over Japan, leading one general to proclaim that he would not trade the 'wonders of paradise for the Kitayama garden' that Yoshimitsu had constructed.

ABOVE: *Ashikaga Yoshimitsu, the third Ashikaga shogun, won all of his military campaigns against the Southern Court, which surrendered to him, and his shugo rivals. He became known as the 'King of Japan' and late in life, depicted here, he acted as if he were a retired emperor.*

LEFT: *Ashikaga Yoshimochi (1386–1428), the son of Yoshimitsu, was appointed shogun in 1394, but held little power until his father died in 1408. Yoshimitsu had been acting like a retired emperor, and once he passed away, Yoshimochi engaged in a reverse course, whereby he emphasized the warrior character of the Ashikaga shoguns. He nevertheless was an important member of the court, and is depicted here in court robes. He frequently consulted with powerful shugo in determining policies and died without an heir, deciding instead to have succession determined by drawing lots.*

The Ashikaga became, as a result of Yoshimitsu's success, remote figures who overawed members of the court and the generals of the provinces. Yoshimitsu meddled in succession disputes, and attacked two of the most powerful *daimyō* – the Yamana in 1392, reducing their 11 provinces to two, and likewise the Ōuchi had their wings clipped after a disturbance in 1399. Yoshimitsu ruled more like an emperor, being involved in lavish rituals, and engaging in intricate prayers, and so he became a remote figure to all but the greatest courtiers.

Yoshimitsu had one son appointed as shogun, named Yoshimochi (1386–1428), who was but a figurehead. Even though Yoshimitsu acted more like a courtier, and at times a charismatic Buddhist priest, than a military figure, he still remained firmly in control. However, his sudden death meant that his resentful son Yoshimochi now wielded power. Yoshimochi systematically overturned nearly all of the policies of his father, abandoning a courtier style, favouring consultation with *daimyō*, and killing his brother Yoshitsugu in the process. Yoshimochi ruled by engaging in consensus with the leading *shugo* of his day, most notably the Hatakeyama, Shiba and Hosokawa. He chose not to command as much as consult and his rule progressed uneventfully.

ASHIKAGA RULERS: YOSHIMOCHI

Yoshimochi strove to be inscrutable, and he succeeded admirably in this effort. He successfully saw off some minor rebellions, and ensured that the Ashikaga would encounter few difficulties during his rule. His only son died young, leaving Yoshimochi with no obvious heir, so in a move that epitomized his style of rulership, he chose not to decide an heir. This led to the *kanrei*, in consultation with a high-ranking priest, to pray at a shrine and to pick the next shogun's name by lottery. One of Yoshimochi's brothers, who served as a high-ranking Tendai priest, was selected.

YOSHINORI

The sixth shogun, known to posterity as Yoshinori (1394–1441), initially took the name Yoshinobu, but this name sounded the same as the term 'the world endures' (*yo shinobu*) which caused the enraged shogun to change his name. Insecurity, coupled with brute violence, characterized Yoshinori's tenure. He executed one of his chefs for his poor

The Golden Pavilion (Kinkakuji) served as a monument displaying Ashikaga Yoshimitsu's power. He built it next to the garden of the Saionji, the most powerful courtier family of the thirteenth and fourteenth centuries, and its location and majesty symbolized Yoshimitsu's fusion of courtly and warrior styles.

Insubordination proved even greater somewhat later, for one *shugo*, fearing for his life, invited Yoshinori to a *nō* play, a highly abstract performance art famously patronized by Ashikaga Yoshimitsu and his successors. During the performance his men assassinated Yoshinori.

YOSHIMASA

The last noteworthy Ashikaga shogun, Yoshimasa (1436–90), strove not to incense anyone, and he ended up annoying all. He vacillated in favouring one man over another, as in the case of the Hatakeyama, and in the end he upset them both. Under his watch, the capital witnessed the Ōnin conflagration, but so great was Ashikaga authority that Yoshimasa himself was not attacked. He preferred to construct Ginkakuji, which is inaccurately known as the 'Silver Pavilion', a temple that was covered in black lacquer instead of gold. Lacquer whitens over the centuries, which is why it later had a silvery appearance, giving it its more recent better-known name.

YOSHITANE

Yoshimasa's son drank himself to death at a young age, and his nephew, Yoshitane (1466–1523), became the next shogun. Yoshitane was ousted in a *coup d'état* in 1493, which was engineered by the *kanrei* Hosokawa Masamoto (1466–1507), and this caused the 'Warring States' era to begin. Yoshitane himself would wander throughout Japan, and ultimately secure the backing of the Ōuchi, returning to the capital from 1508 until 1518 before again being forced to flee.

cooking, and meddled in various inheritance disputes, and at times assassinated prominent *shugo*. The autocrat, terrorizing all about him, caused many of the disturbances of this time. Such erratic behaviour was not enhanced by the fact that Yoshinori drew lots at random in order to decide policy, so great was his faith in this method of decision making. Having served as the head of Enryakuji (Mt Hiei), the main Tendai temple located to the north-east of the capital, Yoshinori had little toleration for religious protest. Enryakuji monks often engaged in violent protests where they would take a holy palanquin and place it in busy intersections in the capital, thereby blocking traffic. Yoshinori demanded that these protests stop, and he commanded his samurai to shoot at these marauding monks, but instead they fled, fearing divine retribution.

The Daikokudō subtemple of Enryakuji, located at the top of Mt. Hiei. Enryakuji priests engaged in frequent armed demonstrations. Battles were waged near the temple in 1336, but its buildings were unscathed until 1571, when Oda Nobunaga incinerated nearly all structures on the mountain. They were rebuilt from the time of Toyotomi Hideyoshi.

The example of the Ashikaga reveals that a cultural shift in ideals of leadership occurred over the course of the fifteenth and sixteenth centuries. The Ashikaga model of rulership became ineffective, and a more active approach came to predominate. This is perhaps epitomized by a contrasting episode concerning Enryakuji, occurring some 150 years after Yoshinori. On 21 September 1571, the warlord Oda Nobunaga ordered his soldiers to burn Enryakuji, killing every monk on Mt Hiei and burning every building, they willingly did so. Warriors were becoming better at following orders as time progressed, and at the same time, seem to have had less fear of any otherworldly retribution.

CHANGING ATTITUDES

A cultural shift in norms of generalship proved great, and explains to a degree why the only generals who were remembered, or idealized in later centuries, were those who died in battle. Among them Nitta Yoshisada and Kusunoki Masashige, each of whom fought against the Ashikaga, became exemplars even though neither proved particularly successful. Masashige had a gift for defending his castle, Chihaya, against Aso Harutoki and the Kamakura armies in 1333, but he had his army annihilated in the fifth month of 1336. Nitta Yoshisada had an even more ignoble demise, in that he blundered into a muddy rice paddy with a squad of cavalry and was killed by a stray arrow.

Yoshisada became, however, an exemplar of 'fair play' – for example, he refused to burn a bridge and thwart the pursuing Ashikaga; or for that matter, challenge Ashikaga Takauji to a duel, with the winner gaining hegemony over the country. Whether such stories are true matter less than that is how they have been remembered.

In death, Masashige became a paragon of absolute loyalty. He was portrayed in a fourteenth-century chronicle, the *Taiheiki* (*Record of a Great Pacification*) as saying that he wished that he would be reborn seven times so as to fight for the emperor. This phrase became manna for nineteenth- and twentieth-century Japanese thinkers who inculcated the notion of extreme willingness to sacrifice for the state. This again reflects a change in beliefs, for in the fourteenth century such statements would consign a figure to hell, for thoughts of salvation and detachment from the world were thought to be preferable to extreme attachment to politics. Unsurprisingly, in the *Taiheiki*, Masashige later appears as a demonic presence. Even in the sixteenth century, however, this devotion to a political cause became the greatest virtue, and Kusunoki Masashige became an increasingly revered figure, with his descendents even being officially pardoned for their earlier 'crimes' against the Ashikaga. The charismatic warlords of the sixteenth century would

NITTA YOSHISADA

Ashikakga Takauji's rival, Yoshisada destroyed the Kamakura regime, but he repeatedly lost out to his Ashikaga rivals and was killed after he stumbled into a muddy rice paddy, and his horse was trapped, allowing archers to kill him. The illustration here shows how fourteenth-century helmets had changed, in favour of a *maidate* decoration instead of earlier *kuwagata* horns. Note too the cloth identification (*shirushi*) on his right sleeve.

take Masashige or Yoshisada as their exemplars. Uesugi Kenshin, for example, would abide by a sense of fairness exemplified by Nitta Yoshisada, even if it proved to be militarily disadvantageous, in that he did not blockade shipments of salt to his landlocked rival, Takeda Shingen, for he did not think that such a victory was worthy of a great general.

Ironically, most of the famous generals of the sixteenth century were, with few notable exceptions, not

The Ginkakuji (Silver Pavillion) was constructed by the eighth shogun, Yoshimasa. It epitomizes the Higashiyama ('Eastern Mountain') culture that coincided with the Ōnin War. This wooden building was originally coated in black lacquer, but centuries of age and exposure to the elements caused this lacquer to turn white (still visible on the window frames), which led to some visitors of the eighteenth century to assume incorrectly that the building was coated in silver.

particularly astute military commanders. Takeda Shingen and Uesugi Kenshin, two of the most noted rivals, lost a large number of battles. The only exception, Oda Nobunaga, had if anything the worst reputation among the 'Warring States' leaders, but he shall be covered in greater depth later in the book.

Hatakeyama Yasaburō, the commander who fought with units of pikemen in Kawachi and Kii provinces, alone appears to have been a tactical genius, for he adopted revolutionary tactics while fighting for his life, but his sudden death in 1459, seven years before the outbreak of the Ōnin War, has caused him to remain unknown.

NEW PATTERNS OF COMMAND AUTHORITY

Generals became famous for their appearance, their personality and charisma, but the nature of tactics in the aftermath of the Ōnin War have

COMMANDERS

standing armies, and troops capable of moving in formation, the role of the commander became significant. The period after 1477 is commonly known as the 'Warring States' period, lasting until 1588 when the samurai order lost its ties to the land, or 1590 when the last *daimyō* was destroyed in battle. These years witnessed the rise of the commander, one of several charismatic figures, wearing flamboyant armour, who were capable of organizing and leading armies.

The role of commanders became significant, for they served to unify their armies. Their sudden or unfortunate death on the battlefield could have cataclysmic consequences. Imagawa Yoshimoto, for example, was of a family that had ruled provinces of east-central Japan for over two centuries. Leading a mighty force at the battle of Okehazama in 1560, he was surprised by troops from Oda Nobunaga's forces and killed. Thereupon his army disintegrated, as too did Imagawa control over the provinces that they had ruled for centuries.

Commanders mattered because they symbolized authority. Most adopted the term *kubō* which means 'Mr Public' and referred to their authority as the 'common good', which meant that they and their lineage personified their domains. They did little actively to engage or command units, which were too large to see a commander's fan, a small foldable object decorated with a

precluded commanders from gaining fame through dashing military offensives. Instead, the ability to fortify and hold a specific area, and amass as large an army as possible, constituted the basis for success. Just as few remember the generals of World War I, and likewise the commanders of the Ōnin War have languished in obscurity.

Nevertheless, Hatakeyama Yasaburō unleashed processes that would lead to profound changes. With the rise of

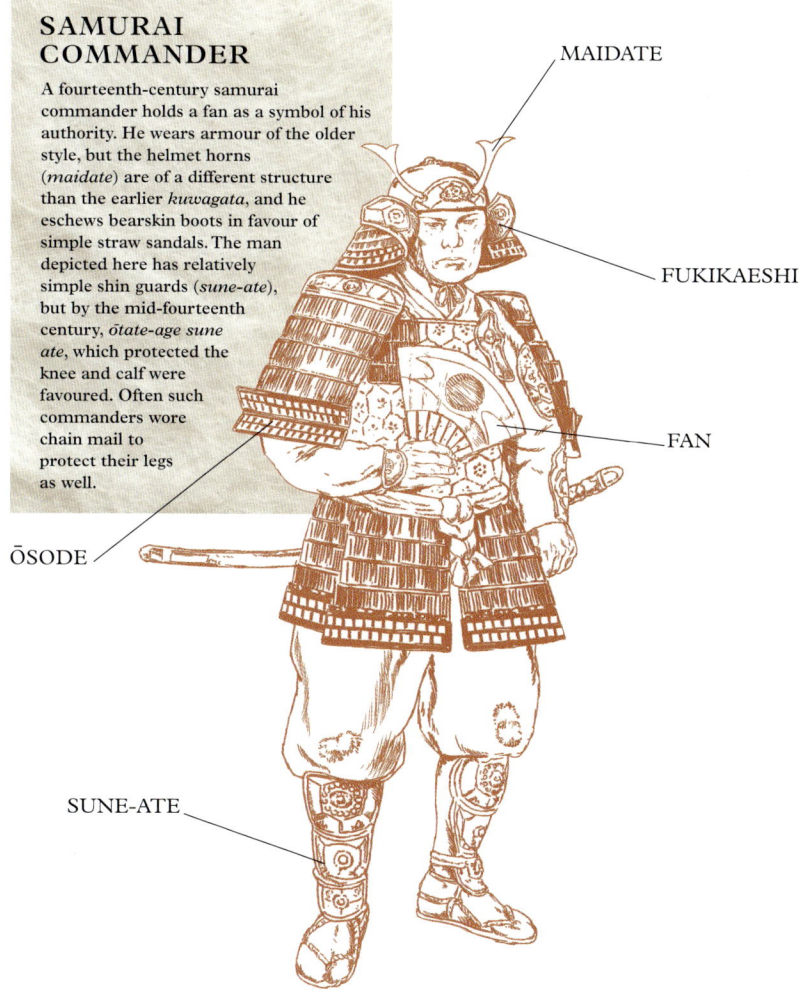

SAMURAI COMMANDER

A fourteenth-century samurai commander holds a fan as a symbol of his authority. He wears armour of the older style, but the helmet horns (*maidate*) are of a different structure than the earlier *kuwagata*, and he eschews bearskin boots in favour of simple straw sandals. The man depicted here has relatively simple shin guards (*sune-ate*), but by the mid-fourteenth century, *ōtate-age sune ate*, which protected the knee and calf were favoured. Often such commanders wore chain mail to protect their legs as well.

MAIDATE
FUKIKAESHI
FAN
ŌSODE
SUNE-ATE

115

The grave of Imagawa Yoshimoto (1519–60), a noted daimyō *who was surprised and killed in his encampment after he thought that he had already won the battle of Okehazama against Oda Nobunaga. Upon his death, his army disintegrated, and the Imagawa, who had been* shugo *since the fourteenth century, collapsed. Nobunaga often carried Yoshimoto's sword with him, and on its* tang *he had engraved: 'The sword of Imagawa Yoshimoto, who was careless and killed by Nobunaga on 19.5.1560.'*

distinct image, but their deaths proved so unnerving that armies would disintegrate, as in the case of the Imagawa army in 1560. In order to better understand the nature of authority, let us turn to the commanders of the 'Warring States' period, for many of them, unlike the Ashikaga shoguns, have been remembered.

Men appointed to the position of *shugo*, constables or lords of their provinces, possessed great advantages. *Shugo*, or *daimyō* as they came to be called, mostly resided in the capital, delegating daily responsibilities to their deputies. Commanders and regional warlords were exceptionally wealthy men, for their positions allowed them to possess half of a province's wealth in taxes. In the aftermath of the Ōnin War, however, authority had to be commanded and rights to the land asserted. Regional domains began to become congruent to lands that could be specifically controlled, and this meant that a greater percentage of the populace had to be mobilized and incorporated into armies, and likewise that specific regions be fortified. In order to better understand the material constraints on these armies, let us explore how they were organized, equipped and protected

SIXTEENTH-CENTURY ARMIES

A variety of mobilization techniques existed, with some successful *daimyō*, such as the Uesugi, relying on their followers as had been typical for centuries before, but they did not take steps to rationalize military service, or the number of troops provided, in relation to the amount of land held. The Latter Hōjō, who ruled eastern Japan from 1495–1590, by contrast proved highly effective in basing military service specifically upon the size of lands held. They effectively broke up the holdings of their followers and made a direct correlation between the amounts of lands held with the number of troops expected to serve.

They also conducted extensive surveys, and valued lands in terms of how much value they had in terms of production. Administrative efficacy did not, however, invariably translate into survivability, as the more 'primitive' Uesugi survived the turmoil of the 'Warring States' period while the more organized Hōjō did not.

CASTLE TECHNOLOGY

Castle technology changed dramatically as well, and the career of the Hōjō epitomizes this. Sōun resided in the mountainous Nirayama Castle in Izu. He constructed this castle by using a rope and laying it along a hilltop, tighting it on stakes. This 'pulled rope' (*nawabari*) became the name for this type of castle, for the areas enclosed by the rope would become earthen walls that were built up so steeply that plants could not readily grow on them. At times, panels of wood would temporarily cover these walls so as to give them shape. Above these walls, wooden structures were added, while next to them, large moats were excavated. Starting in the 1540s in central Japan, weak spots in earthwork walls would be reinforced with a few boulders – this is evident in castles such as Shōrakuji, located in Ōmi province to the east of the capital. These central regions lagged behind in castle technology, however, for masons in western Japan proved far more skilled, constructing straight and formidable stone walls as early as 1508. Castles with stone walls can be verified as early as the 1560s, such as Kaizu, a castle built by Takeda Shingen near the Kawanakajima battleground.

By the 1570s nearly all castles would be built on plains, in central locations for markets, rather than in easily defensible mountains. Improvements in building techniques caused commanders to have more faith in their stone walls, but this defensive emphasis itself paled in comparison with the need to control markets for trade. Taxes, trade and access to markets mattered as much, if not more, than the ability to defend a particular area. This need becomes all the more clear when exploring how armies armed their men.

SIXTEENTH-CENTURY ARMOUR

Regardless of their organization, all armies required larger numbers of troops, which entailed changes in the arms and armour of the samurai. One begins to see something approaching uniforms in the sixteenth century, with armour being marked with family crests of *daimyō*. Particularly noteworthy are some surviving suits of armour of the Latter Hōjō, who used a distinctive triangular crest from 1524 through 1590. This armour had the *daimyō*'s crest emblazoned on the torso so as to make it

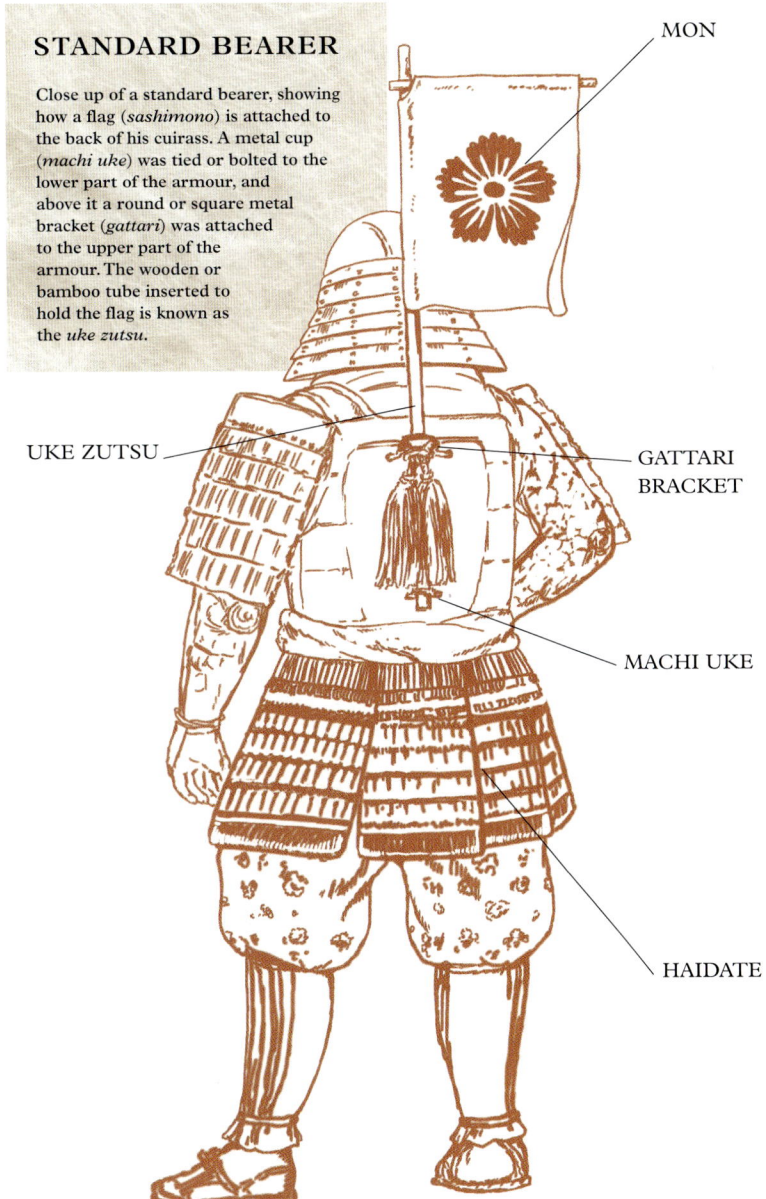

STANDARD BEARER

Close up of a standard bearer, showing how a flag (*sashimono*) is attached to the back of his cuirass. A metal cup (*machi uke*) was tied or bolted to the lower part of the armour, and above it a round or square metal bracket (*gattari*) was attached to the upper part of the armour. The wooden or bamboo tube inserted to hold the flag is known as the *uke zutsu*.

- MON
- UKE ZUTSU
- GATTARI BRACKET
- MACHI UKE
- HAIDATE

TŌSEI GUSOKU ARMOUR

Once helmets were created in the sixteenth century that could protect the head with just a few plates of metal, warriors took advantage of this simplified construction to add elaborate decorations so as to make them, particularly commanders, more recognizable on the battlefield. These helmets, known as *kawari bachi*, 'unique' or 'different' helmet, epitomize the flamboyance of the late sixteenth century. In contrast to the deer horns of Honda Toshiaki, here one sees giant oxe horns and a tassle attached to the top of this seventeenth-century example. This suit of armour also uses more lace, and has some elements, such as the *fukikaeshi* on either side of the helmet, which hark back to earlier styles.

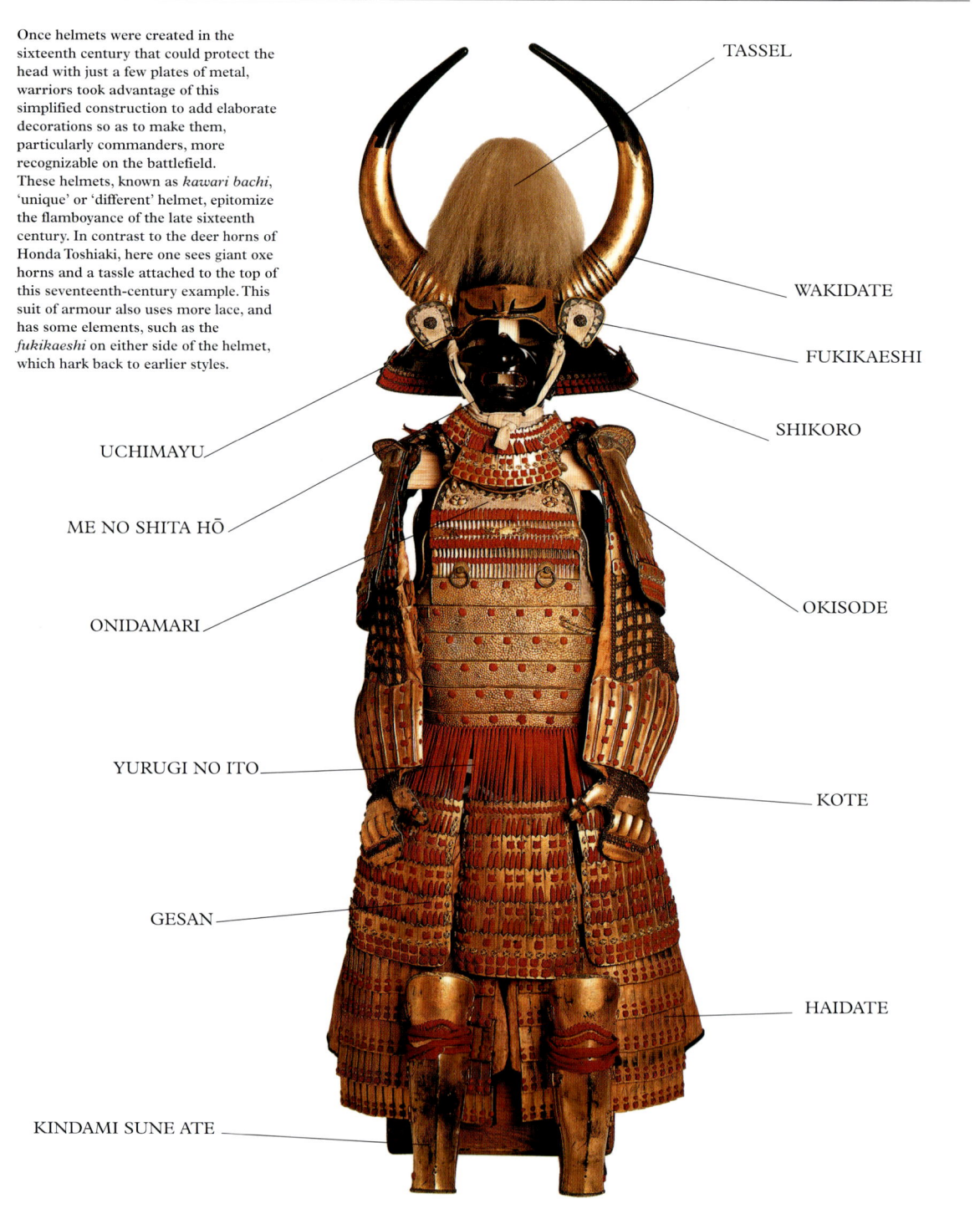

PAPER ARMOUR

Creating professional armour for every soldier proved problematic, and so some individuals relied on, of all things, paper armour. One suit, owned by Izawa Shōji, is made entirely of Japanese paper, and although one might think that such material is inappropriate for protection, in fact this proved adequate. Japanese paper is far stronger and more durable than paper from Europe, for it is made from mulberry fibres and is virtually indestructible. This paper would be covered in lacquer and so represented a hard, light suit that was impervious to cracking, unlike leather, and provided equal protection. Lightness proved a particular advantage, and even the shin guards and gloves (*kote*) were made from paper. The above helmet is made of lacquered paper and had the advantage of being light, rust proof and providing a modicum of protection.

STYLES OF MAIL

Early Japanese mail was used from the fourteenth century onwards, but the oldest surviving examples are from the sixteenth century, when chain mail became very popular and its techniques of construction improved. A variety of weaves exist, which are counted according to how many times each chain was linked – be it three, four, six or eight times. Chain was used for shin guards, gauntlets, leg protection and neck guards. Here are some of the weaves. *Nawame-gusari* represents an older way of linking individual links of mail. *Nanban-gusari* owes its name to Europeans, or 'southern barbarians'.

readily identifiable. The terminology for sixteenth-century armour begins to shift in that it is referred to as 'current armour' (*tōsei gusoku*) and it varies considerably from earlier styles. The word *gusoku* means 'these are adequate' and generally denotes a helmet plus metal armour protecting the torso, and arm and leg protection. The term itself is old, dating from the thirteenth century, but became more prominent in later centuries to describe armour.

TORSO PROTECTION

The later styles of armour were better at protecting the body with few, if any, openings in the torso section. Metal became more favoured than lacquer, although lacquered panels remained. Nevertheless, one major change was that chain was at times used to link panels of metal together, and braiding changed to a style called *sugake*, which meant 'simple hang', which replaced the earlier style of braiding, known as *kebiki* or 'pulled hair'.

The *sugake* style entailed using less braid – this proved advantageous, for braiding could become encrusted with dirt or lice, and took time to dry out from rain in long campaigns. Further compounding difficulties, waterlogged lacing could freeze, making braided armour unbearable in wintertime. Some commentators, such as Sakaibara Kōzan, have postulated that braid also served to trap a pike tip instead of letting the blade glide off harmlessly. The sixteenth-century suit of armour worn by the female warrior Tsuruhime exhibits this distinctive x-shaped *sugake* weave.

Another type of armour, known as *karuta gane tatami dō*, consisted of small rectangular plates of armour – which resembled cards (*karuta*) and hence were called 'card metal' (*karuta gane*) sewn armor (*tatami dō*) – and chain sewed onto fabric backing, thereby dispensing with the need for lacing altogether. This armour proved to be inexpensive and durable and became the favoured armour for lower-ranking soldiers. In short, changes in armour contributed to its durability in the field, and helped to protect against pikes.

The Latter Hōjō adopted, as we have seen, a sophisticated and comprehensive military organization, which involved an immense number of troops. The Hōjō were able to put their crest on nearly all suits of armour, thereby making all of their warriors readily identifiable.

Typical is one suit owned by Nakajima Kenki, which consists of *dōmaru* armour made of a reinforced metal torso guard, with metal plates covered with leather and the lower body protected with plates woven together, much as had been the case for centuries albeit with a simplified torso section with no braiding save for some on the lower torso. Prominently displayed in the centre of the chest is the Hōjō crest emblazoned in gold.

HELMET STYLES

The helmet, too, changed radically over the course of the fifteenth and sixteenth centuries. Yamakami Hachirō's *Katchū no*

Yuki no shita dō. *This cuirass is of the style of Yuki no shita armourers, who resided in Kamakura. This style was used by warriors who fought in the armies of the Latter Hōjō.*

shin kenkyū reveals that as early as 1510 a new type of helmet arose, which was crafted by Myōchin Nobuie. The helmet is made of simple iron with no decoration, the eight plates of finely crafted metal held in place by metal rivets. Other helmets favoured decorative rivets, with some having as many as 2000, which did nothing for the helmet save add weight to it, but this proved a popular decoration. Most were made of steel, but some were also made from hardened leather.

Another new type of helmet, called *zunari*, which means 'shaping the head', consisted of a small number of shaped plates, and it could be of heavier steel because of its simplicity, and thus offer

ZUNARI HELMET

Zunari, or 'head-shaped' helmets, combined enhanced protection with thick steel plates and a simplified design. The *shikoro* hangs from this helmet and protected the back of the neck. This construction proved so durable that elaborate decorations could be added, although, as in this example, some preferred an extremely simple design, with the only decoration being that of a small lobster head and red and gold lacquer.

SIXTEENTH-CENTURY HELMETS

The sixteenth century witnessed a revolution in helmet design. The central plate of the armour, covering the crown, overlapped the plate covering the brow. This allowed *zunari* helmets to be made from fewer plates of heavier steel. As these effective helmets were rather plain, soon warriors personalised their helmets with elaborate decorations.

See the 'peach-shaped' helmets (*momonari-kabuto*) which are emblematic of this style. *Tōjin gashira* are based on Chinese or Mongol style helmets, with a Japanese-style *shikoro* hanging from it, while the *tori*, or bird helmet is decorated to look like a bird. *Hoshi-kabuto*, or star helmets, are in the style of older helmets, with the rivets here function as decoration. *Jingasa* helmets, used mostly for foot soldiers, are first documented in 1575, and appear to have initially been used in eastern Japan. This type of protection became common for low-ranking warriors and firemen in the seventeenth, eighteenth and early nineteenth centuries.

KABUTO

MOMONARI-KABUTO

TŌJIN GASHIRA

MOMONARI-KABUTO

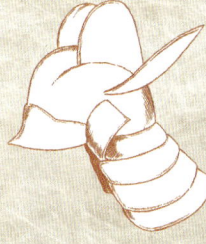
TORI-KABUTO

HOSHI-KABUTO

HOSHI-KABUTO

Ō-BASHI KABUTO

JINGASA (WIDE-BRIMMED)

JINGASA

JINGASA

greater protection. These helmets were named after families who prominently used them, the Hineno of Izumi, to the south of the capital, and also the Etchū, which referred to a member of the Hosokawa family who favoured this style. Accompanying this style, craftsmen created more elaborate face guards, many shaped like an actual face, which protected greater protection than earlier versions.

Once armour makers could construct helmets from a few metal plates, new and innovative designs emerged. Generals and warriors competed, or so it seemed, in wearing novel helmets or *kawaribachi*. Some were created that resembled *eboshi* while other looked like catfish tails. Still others resembled Chinese-style caps, a cresting wave, the tail of a mackerel, a conch shell, the horns of deer, and one of the most unusual examples resembles the face of a monkey. As larger armies came to exist, commanders began to rely on novel or unique styles or armour, or in particular helmets, to stand out from their troops. The helmet and armour of most warriors, in contrast to the flamboyance of the

> ## SUGAKE ARMOUR
>
> *Sugake* or 'simple hang' represents a style where armour was laced vertically. This drawing is based on the armour of Honda Tadakatsu (1548–1610), one of Tokugawa Ieyasu's most trusted generals, and a *daimyō* of Ise Province. The leering face (*shigami*) at the front of the helmet was a popular design, and the antlers here were made of wooden and lacquered paper. Although difficult to see, the leggings and arm protection were made almost completely of chain mail with some reinforced metal strips, and the shin guards (*sune-ate*) are likewise mostly made of chain. The rosary, or *kinpaku oshi juzu*, draped around his armour was a sign of his Buddhist devotion.

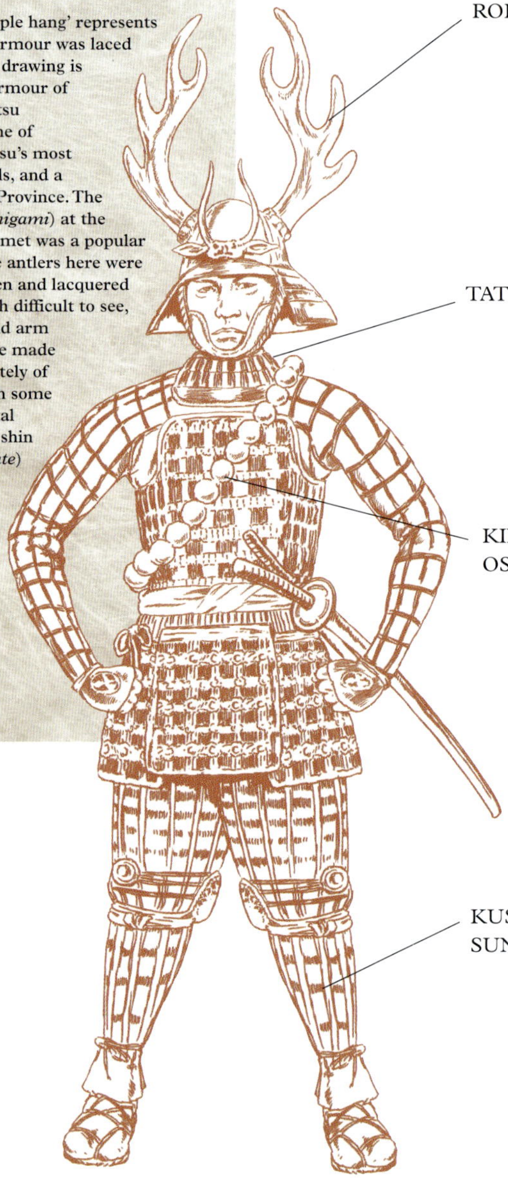

ROKKAKU

TATE ERI

KINPAKU OSHI JUZU

KUSARI SUNE-ATE

commander, showed less status distinctions than had been common before. A variety of armour makers became prominent, such as the Haruta of Nara and the Iwai, both of whom

Tokugawa Ieyasu wearing a European cuirass combined with Japanese-style skirt and extensive mail protecting his arms and legs.

started signing their armour. Another school, the Myōchin, also arose and was based in Sakai. This school is famous for the high quality of its armour, and the earliest reliable work attributed to its founder, Myōchin Nobuie, consists of a helmet dating from 1510, while the last surviving example of his work dates from 1544. Nevertheless, Nobuie became so famous that earlier examples of armor,

COMMANDERS

Menoshita *faceguards were common in the late sixteenth century, which protected the face below the eyes and the neck. This provided more protection than the earlier* hō-ate *and could represent human, animal, or goblin faces, thereby appealing to the individuality of the wearer and at the same time serving to intimidate opponents.*

some created as early as 1472, were anachronistically attributed to him and his school. Schools of armor makers arose in other regions as well, such as Saotome group, prominent in Hitachi Province in eastern Japan, and the Yuki no shita, who resided in Kamakura and made armour for the Latter Hōjō. Arms manufacturers became more dispersed and tended to become more associated with particular *daimyō*.

Over the course of the sixteenth century, heavier steel armour came to be favoured, with the torso benefiting from the greatest increase in protection. Plate armour, consisting of seven or eight vertical metal plates welded together (called *yokohagidō*), was developed in the 1540s in Owari Province. Other versions, with horizontal plates (*tatehagidō*) exist as well, while the Yuki no shita smiths manufactured metal breastplates that came to be known as *Yuki no shita dō* for the Hōjō armies.

Tokugawa Ieyasu took the breastplate of some Spanish armour and incorporated Japanese-style accoutrements, and these became known as 'southern barbarian breastplates' (*nanban dō*). This was followed by large single pieces of reinforced steel, known as 'one-piece armour' (*ichimaibari uchidashidō*), which consisted of one strengthened plate of steel in the front and one in the back to provide greater protection from bullets (for more on Japanese guns, see Chapter 5).

THE VARYING STRATEGIES OF JAPANESE WARLORDS

Commanders often faced difficulties because they possessed nominal authority over several provinces, having delegated their authority to deputies. The deputies were well aware of the situation in the provinces, and able to wield the machinery of government. Preoccupied or unskilled commanders were liable to lose their positions to ambitious deputies. Some very prominent families, such as the Yamana and Akamatsu, rapidly withered to irrelevance. Others, such as the Hosokawa, had more staying power until their lord, Masamoto, was assassinated by three of his followers while taking a bath in 1507. A few, most particularly the Imagawa, Ōuchi, Takeda, Uesugi and Shimazu, proved successful in maintaining their authority, at least through the mid-sixteenth century. Rather than relying on generalizations, we can explore a few of these warlords in turn.

THE CONSERVATIVE WAR LORDS: THE ŌUCHI

The Ōuchi proved to be amongst the most powerful *daimyō* of the fifteenth and early sixteenth centuries. Ōuchi

*In contrast to the simplicity of most sixteenth-century helmets (*zunari*), some preferred the opposite approach. This helmet has 62 ribbed lines of rivets (literally stars or* hoshi*) attached, while the* maidate *helmet decoration is of an off-centred crescent moon.*

123

TŌSEI GUSOKU ARMOUR

Tōsei gusoku represents a new type of armour, devised in the late sixteenth century, that was lighter and more flexible, and relied on less lacing than earlier armour. It also provided greater protection for more of the body, so as to prevent pike wounds. In particular, protection for the lower abdomen was enhanced. Empty suits of armour were strong enough to sit upright, which is why sometimes these suits are known as 'standing armour' (*tachi dō*). Most suits were made of two pieces, known as *nimaidō*, which hinged under the left arm. When plates known as *iyozane*, were laced vertically (a style known as *sugake*), the armour is called *nuinobe-dō*, which means 'extended weaving'.

ME NO SHITA HŌ

ONI DAMARI

KUSARI BAKAMA (CHAIN MAIL LEGGINGS)

Masahiro's intervention proved crucial for the continuation of the Ōnin War, and after the war ended Masahiro returned to western Japan, where the created his own model city of Kyoto, with similarly named temples and rituals. The town of Yamaguchi looked more imposing than the capital after 1477, with five-storey pagodas built for the Rurikōji Temple, which still survives as a testament to Ōuchi wealth and power.

The Ōuchi region of Yamaguchi witnessed technological progress, in that its stonemasons proved able to build level fortified walls far earlier than could be done in central Japan, which had to wait until the 1540s. The Ōuchi were also deeply involved in trade with China, and had close ties to Korea. They even declared themselves to be 'sons of shooting stars' who were reborn as a Korean prince who then settled in Japan. Their lineage proved far less illustrious, for they stemmed from pedestrian warriors of the thirteenth century, but their ties to Asia led some to call them 'Mongols and not Japanese'.

LIMITED VISION

The Ōuchi, being combatants in the Ōnin War, fought using current military tactics, relying heavily on pikes just as their contemporaries, but their methods of mobilization more closely resembled fourteenth- than fifteenth-century patterns. Ōuchi warriors retained more autonomy, and they continued to submit petitions demanding rewards for service, something that was typical for warriors of the fourteenth, rather than the fifteenth, century. Having greater stability than nearly every other region, the Ōuchi were able to keep their deputies in line initially. The Ōuchi stronghold also represented a natural place for Yoshitane, the Ashikaga shogun who was ousted by Hosokawa Masamoto's coup in 1493, to flee. With the greatest army, the Ōuchi were well placed to transform Japan if they so wished, but instead they used their military resources to uphold the Ashikaga, a trait that they shared with another notable leader, Uesugi Kenshin.

LEFT: *Rurikōji. A five-storey pagoda constructed by the Ōuchi in 1442. The Ōuchi modelled their town of Yamaguchi and its temples and shrines on Kyoto. After the Ōnin War, this 'Little Kyoto' attracted artisans and nobles from the burned-out capital.*

RIGHT: *The banner of Ōuchi Yoshitaka, with the daimyō's family crest and also the names of various protective deities.*

Their conservative expectations for their followers mirrored the limits of their vision. Once their arch-rival, Hosokawa Masamoto, who had alienated his followers through erratic behaviour, including acting like a Buddhist priest and performing curses, and prohibiting all women in his presence, was killed in 1507 leaving no undisputed heir, the Ōuchi led a mighty army to restore the Ashikaga shogun. It is a sign of their success that they managed to do so for 10 years, from 1508 until 1518, for there they remained in the capital, and the Ashikaga witnessed a partial revival. But in 1518, Ōuchi Yoshioki feared losing control of his deputies, and was in danger of being worsted in battle by his rivals, so he left the capital. Ashikaga Yoshitane fled as well and wandered about Japan before he perished.

The Ōuchi pattern of rulership resembled that of the Ashikaga in that they cultivated poetry and acted like a capital aristocrat, going so far as to model their homelands in Yamaguchi on the capital of Kyoto. They also proved effective in regularizing exchange rates and crafting laws after the Ōnin War. These measures limited violent crimes and encouraged individuals to pay their taxes. With the stability of their domains they were able to patronize local temples and shrines, and even construct lavish residences and temples, most notably the five-storey pagoda at Rurikōji. Some of their temples, such as at Ryōunji, are remarkable in that they have well-crafted stone walls, dating to 1507, which are 60m (197ft) long, over 3m (9ft 10in) high, and over 2m (6ft 6in) wide. This was constructed under the aegis of Ōuchi Yoshioki (1477–1528), the thirteenth leader of the domain. These stone walls

WEAPONS AND FIGHTING TECHNIQUES OF THE SAMURAI WARRIOR

A sixteenth-century depiction of armour makers. Most continued to reside in the capital of Kyoto, and their stores and dwellings, located in the south of the capital, were not damaged during the Ōnin War.

are of greater quality and sophistication than the earthen fortifications that existed in central Japan at that time.

Yoshioki ruled his domains from 1494, after his father Masahiro retired, until his death in 1528. He was successful in almost all of his endeavours, until he was defeated by rival *daimyō*, the Amako, in 1521. Yoshioki managed to soundly defeat the Amako three years later, but he could not destroy them. His son Yoshitaka (1507–51) continued fighting the Amako, and he notably routed them in 1539, but like his father he could not decisively destroy his rival, and he endured a loss in 1542. A dispute arose among Ōuchi retainers as to whether to expand the conflict, or prefer negotiations. The Sue, who favoured a vigorous military response, grew increasingly dissatisfied with their Ōuchi lords. Defeat caused Ōuchi support to waver in the southern part of their domains, and ultimately his deputy, and most trusted retainer, Sue Takafusa, rebelled against him in 1551, forcing Ōuchi Yoshitaka to flee and ultimately take his own life in the ninth month of 1551.

The Ōuchi proved to be a bastion of stability after the Ōnin War, but they culturally and military modelled themselves after the Ashikaga, going so far as to use most of their military strength to uphold the old order, rather than forge a new one of their own. This cultural conservatism overshadowed genuinely innovative aspects of their regime, such as the employment of sophisticated stone-walled fortifications far earlier than used elsewhere in Japan, and the early receipt of Portuguese guns. Nevertheless, access to these new weapons and techniques did little to further Ōuchi rule. They always looked backward, even when confronted with Christian missionaries they believed that they merely represented a new sect of Buddhist monks coming from India.

ABLE ADMINISTRATORS: THE LATTER HŌJŌ

The Latter Hōjō, another prominent daimyo, proved more innovative than the Ōuchi, but in the end were no more successful. Most histories portray them as being self-made men with their founder, Hōjō Sōun, being a wandering samurai who overthrew others in eastern Japan and became the first new 'Warring States' *daimyō* in 1493. In fact, he was not a humble man of lowly origins, but from the Ise family of administrative specialists for the Ashikaga. Although not of the main Ise line, he still served as a messenger (*mōshitsugishū*) for the *bakufu*, and was a close confidant of Ashikaga Yoshimi, who worked closely with the Western army during the Ōnin War.

Sōun fled from the capital while still a young man, avoiding the Ōnin conflict, and settled in the Imagawa domain, where he solved a succession dispute and then travelled to eastern Japan. When the shogun Yoshitane (who was Yoshimi's son) was exiled following a coup in 1493, and a puppet Ashikaga installed in his place, Sōun acted on behalf of the Imagawa and attacked and killed the pro-

The walls of Ryōunji, built in 1508, reveal how the Ōuchi were capable of building structures made from cut stone. Such skill was unknown in other areas of Japan at the time, for fortifications constructed in eastern and central Japan were still made from earth. Earlier examples of stone walls, such as those built in the 1270s to defend Japan against the Mongols, merely consisted of rocks being piled up and not cut and crafted like this.

Hosokawa Ashikaga representative in eastern Japan in 1493.

Historians have often classified Sōun as epitomizing the new era, of *gekokujō*, or lower conquering the higher, but this emphasis is misplaced. Sōun was related to a prominent family of Ashikaga administrators, and his actions did more to damage Hosokawa interests than further a revolutionary movement to overthrow authority. He acted against the Ashikaga in the east because he did not accept the legitimacy of the Hosokawa coup, an attitude he shared with the Ōuchi in the west at this same time.

That being said, Sōun was remarkable in that he leveraged his position as an Ashikaga official to one where he became a regional lord. According to lore, he reached an agreement with six other samurai, and they agreed to remain together until one of them achieved success. These original 'seven samurai' witnessed one of their members retiring and another taking religious orders, but otherwise the remaining five men, and their descendants served Sōun and his descendants well. Attracting such support helps explain Sōun's success on one level, for any commander required close followers, but at the same time alliances were necessary. Sōun worked completely in tandem with the Imagawa, a neighbouring *shugo* to his west. Practically, if not officially, Sōun behaved as an Imagawa deputy *shugo*, and helped strengthen the Imagawa's position as the dominant warlord family of south-central Japan. Sōun and the Imagawa fought primarily against the Uesugi family, a venerable family who were appointed as the 'deputy shogun of the east' (*kantō kanrei*).

Sōun himself proved wily, for he and his troops were able to capture Odawara Castle by pretending to be a hunting party. The lack of wariness on the part of his opponent, the Uesugi follower Ōmori Ujiyori, stemmed perhaps from the fact that, even as late as 1493, such a brazen attack was inconceivable. Still, throughout his life Sōun remained entrenched in a smallish mountain castle at Nirayama. When fighting he only attacked to the east, for he remained allied to the *daimyō* of the west, the Imagawa, and indeed Sōun, writing as late as 1508, referred to Imagawa Ujichika as his lord (*yakata sama*). He maintained good contacts to the east, and even received a new weapon, the harquebus (*teppō*), in 1510. Not until 1512 did he defeat another rival, the Miura, and advance into Kamakura, site of the first *bakufu*. He wrote a poem at this time that when paraphrased reads as follows: 'Planting a new flower by the withered tree, see how the old capital returns.'

With the capture of Kamakura, Sōun changed his name to Hōjō, and took on an identity as the successor to the able administrators of Kamakura. It took Sōun 20 years to capture the single province of Sagami, but even prior to this success, he began surveying his lands yearly from 1506, and he also penned a simple body of laws, consisting of some 21 codes. Hōjō Sōun, rather than epitomizing the spirit of *gekokujō*, was in fact an able administrator who managed to see the opportunities that post-Ōnin Japan afforded. He worked with the Imagawa, attracted loyal supporters, and through a variety of stratagems, gained local power in eastern Japan, and continued his ongoing conflict with the Uesugi in the east.

CONTINUING CONFLICT
In political terms, Japan after 1493 witnessed a fracture between two major factions, which represented a continuation of the Ōnin War as the fissures of that conflict continued to widen. Hosokawa Katsumoto (1430–73) led the Eastern army, and his son Masamoto engineered a coup against the

A depiction of Takeda Shingen and his retainers. Successful daimyō delegated authority to a group of advisors and generals, and the importance of this group is epitomized in this image. The Takeda retainers became famous for their devotion to the Takeda and their skill in battle. After the Takeda were destroyed, the descendants of many of their followers eventually found service with the Tokugawa, who established their bakufu, *or warrior government, in 1603.*

Ashikaga. Eastern Japan, particularly the Imagawa and their able followers, the Hōjō, and western Japan, under the Ōuchi, did not recognize the legitimacy of the Hosokawa coup, and ceased to obey central authority. By contrast, the Uesugi continued to be tightly linked to the court.

As time passed, however, any inadequacies in command resulted in deputy *shugo* overthrowing their superiors, particularly around the year 1550. If one can see a *gekokujō*, it was merely of deputies taking over for their ineffectual superiors, or for that matter, asserting autonomy from them, as the Latter Hōjō would finally do when Sōun's son, Ujitsuna, captured Edo Castle in 1524. This incident started a major offensive to capture the plains of Musashi province, and attack the Uesugi, another *daimyō* family and a hereditary power in the east. It was at this moment that he changed his name to the Latter Hōjō, so as to legitimate his attack against the Uesugi. In 1524 Ujitsuna adopted the distinctive triangular Hōjō crest, so as to emphasise his position as heir to the Hōjō family, administrators who had dominated the Kamakura *bakufu* from 1200 until 1333. From 1532 Ujitsuna likewise rebuilt the major shrine of Tsurugaoka Hachiman, which had been burned in 1526, again to strengthen his ties with Kamakura. Ujitsuna would fight the Uesugi, who believed themselves to be the most important family in the east, for 17 full years after he captured Edo Castle.

Ujitsuna's autonomy had its limits, for Imagwa Ujichika was Sōun's nephew. As late as 1535, the Hōjō leader still aided the Imagawa against an attack by their Takeda rivals. Two years later, however, the next Imagawa lord, Yoshimoto (1519–60), took a daughter of Takeda Nobutora (1494–1574) as a wife, and started actively fortifying the eastern border between their domains, which had hitherto been vaguely defined. This meant that the Latter Hōjō now became autonomous from the Imagawa, as they were now forced to defend their common border region. Nevertheless, memory of their earlier ties remained, in spite of the Imagawa alliance with the Takeda, and ultimately, in 1554, the Imagawa, Takeda and Hōjō would all forge an alliance.

Sōun passed his chieftainship to his son Uchitsuna in 1518, and passed away the following year at the remarkable age of 88. His son Uchitsuna would die in 1541, but he continued to uphold the Hōjō codes and expand the Hōjō domains. He wrote a five-clause document to his sons, which reveals much about his attitudes. First, he wrote:

the behaviour of even commoners, who would themselves desire to appear wealthier than they were, and thereby rely on gambling, and some samurai, unable to appear ostentatious, would flee. In the end, the lord would prosper but the rest of the domain would not – which is how the Uesugi behaved, Ujitsuna argued. Likewise, Ujitsuna argued that samurai should be treated according to their rank.

Finally, in a closely related point, Ujitsuna argued that frugality above all

TOKUGAWA IEYASU

Tokugawa Ieyasu, depicted here wearing armour, lost battles against Takeda Shingen and Toyotomi Hideyoshi, but he triumphed at Nagashino, and later Sekigahara in 1600. Ieyasu had signed an oath to support the Toyotomi, but after Hideyoshi's death, he violated it, skilfully attacking Hideyoshi's lieutenants in 1600 and ultimately relying on a feeble excuse to destroy Hideyoshi's heir in 1615. Ieyasu's actions engendered resentment, and, even though he founded a powerful shogunate in 1603, his regime relied upon coercive measures and surveillance to maintain power.

> *Generals and samurai should be principled. Without principles, even if one is to take a province or two, in later ages it will be a sources of shame, and if and when the favour of heaven is exhausted, if one knows that principles have not been compromised, then there will be no shame (or criticism) in later generations… Even if one is to read old stories, there is all the difference in the world, as great as between heaven and earth, between those who upheld principles and were destroyed, and those who abandoned principles for prosperity.*

Here, Ujitsuna is criticizing the Ashikaga, who defeated the imperial family, and implicitly favouring their more 'virtuous' rivals, the generals Nitta Yoshisada and Kusunoki Masashige, who established reputations for fidelity, honesty, fairness and bravery.

Ujitsuna's second point was that men of all status, from the highest-ranking warriors to the basest commoner, had abilities, and that it was vital for a general who controlled a province to use all men of that province. Conversely, Ujitsuna realized the limitations of even the most illustrious individual, by stating that no one is extraordinarily able in all endeavours.

Ujitsuna next stated that samurai should not be arrogant, nor toadies, but protect their status. In order to do so, they should not be forced to appear lavish, for this would cause them to levy heavy taxes on peasants and merchants; such a lust for extravagance would affect

should be valued. Here frugality meant the frugality of the lord, for if the lord eschewed flamboyance, and did not overly tax his domains, then all would prosper, warriors and peasants alike would be willing to serve as the point of the spear of the Hōjō armies. In conclusion, he also advised against getting arrogant in victory, or belittling a defeated enemy. Ujitsuna's 1541 testament represents a radically new vision of samurai ideals.

To the Ashikaga or the Ōuchi, frugality represented a foreign notion of governance and behaviour, but these ideals would achieve wide acceptance over the course of the sixteenth century. Another *daimyō*, Asakura Toshikage (1428–81), wrote a code emphasizing frugality, while another member of this family, Norikage (1474–1555), would recount how impressive Sōun was for even carefully collecting stray needles and preserving them in his storehouse.

BATTLE FORMATIONS

As armies became larger and tightly organized, they gained the ability to create tactical formations, with pikemen to the fore protecting the main body of troops. The central region served to protect the commander, for they proved so important that their death would so unnerve an army as to cause it to disintegrate. Most often relatively simple formations like *ganko* were used, but when armies encamped they would create more elaborate positions.

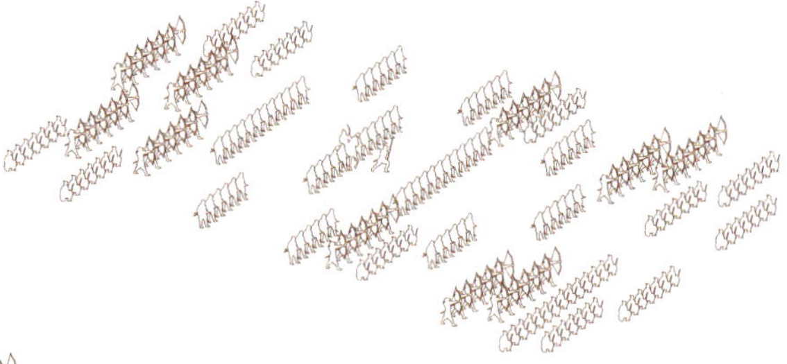

GANKO

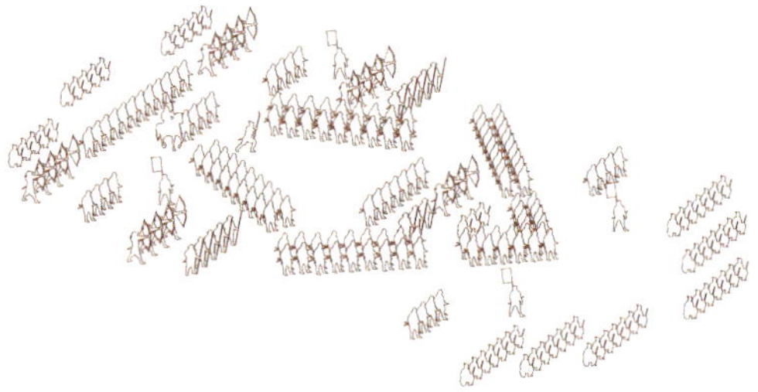

HOSHI

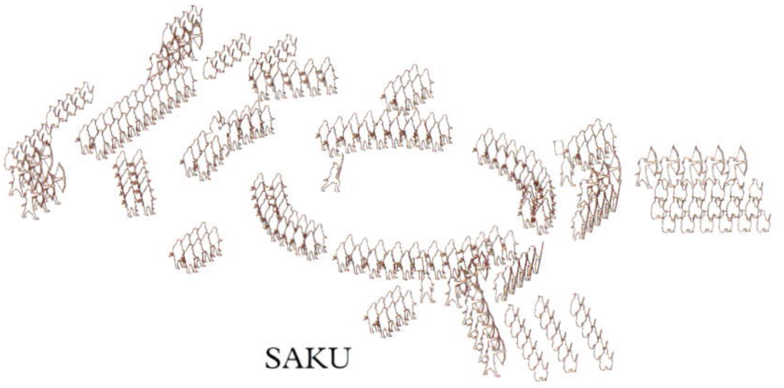

SAKU

KAKUYOKI

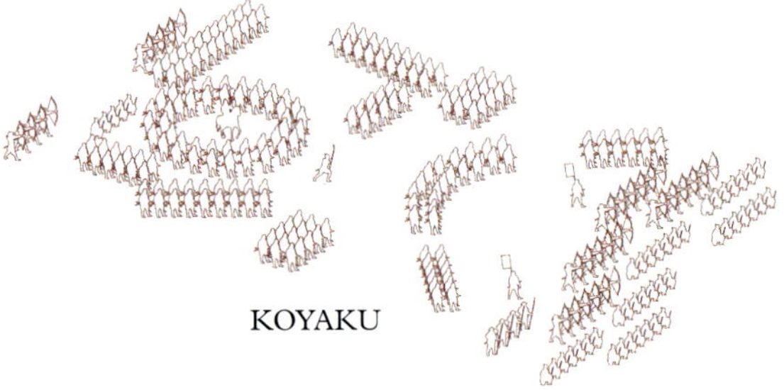

KOYAKU

GYORIN

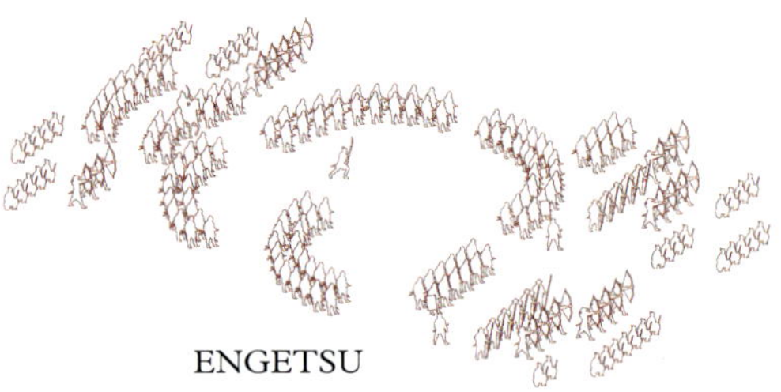

ENGETSU

COMMANDERS

There is a gulf between the actions of Sōun and his successors, most particularly Ujitsuna, and the Ōuchi. Sōun was capable of defeating his rivals through trickery, or long sieges, such as those that allowed him to capture Odawara Castle. Yet having done so, he and his descendants engaged in the most impressive reorganization of their domains, for they surveyed their territories and assessed revenue according to productivity rather than merely the size of the land.

Hōjō samurai lands were divided up and scattered, making any rebellion impossible. Likewise, the Hōjō constructed an impressive network of castles throughout their gradually expanding domain, making any attack extremely difficult. The Hōjō innovated in other ways. Precisely because they maintained a network of castles, they improved communication. Often, signal fires were used to transmit warnings – documented use of such fires in Japan reach back to the Mongol invasions of 1281. Fires could not be easily seen, however, during the day, or readily made in times of rain, and so the Latter Hōjō preferred using musical instruments. They used bells, first, as a command for castles to collect supplies, and then the *taikō* drum for soldiers to put on armour and head out to encampments. Finally, if the solders were ordered to retreat to their castle, a conch shell was sounded. The Hōjō did not reject signal fires, but relied on dried wolf dung apparently to change the colour of smoke and so make the signal fires distinctive. This communication was most readily used during night attacks at sea between the Hōjō and their Satomi rivals.

HŌJŌ MILITARY ORGANIZATION

The research of Michael Birt, who wrote a dissertation on the Hōjō, has revealed how the Hōjō army relied on a decentralized political structure, revolving around 'satellite castle units' that provided for territorial occupation.

The keep of Odawara castle, centre of Hōjō domain, which withstood a siege by Uesugi Kenshin and was thought to be impregnable. In 1590, Toyotomi Hideyoshi surrounded this castle and starved the Hōjō into submission.

The main shrine of Tsuruoka Hachiman, located in Kamakura.

the oldest surviving Hōjō mobilization rosters, dating from the eighth day of the third month of 1556, reveals that Inami warriors with land worth 442.4 *kan* were expected to mobilize 56 men, of a rate of 7.9 *kan* per warrior. Nothing was mentioned as to what weapons these men should carry save that 12 of the 56 were to be horsemen. This figure suggests, likewise, that only one fifth of the Hōjō armies consisted of horsemen, a radical change from earlier armies, and consistent with the post-Ōnin pattern whereby pikemen predominated.

MOBILIZATION REGISTER

A most informative document, dating from 1559, reveals how the Hōjō had created a register, showing that 560 warriors possessed revenue worth 72,168.3 *kan* of cash. We do not know how many men these 560 mobilized, but can still ascertain much about the Hōjō army. First, companies were based on a geographic rather than kin-based distribution. In all probability, as many as 10,000 men could be mobilized according to this register, and as this register did not include all of the Hōjō lands, one can extrapolate that their armies overall consisted of approximately 20,000 men in 1559. The most loyal of the Hōjō retainers were given lands on the periphery of the domains, which constituted an innovative practice, for it ensured that these regions would be firmly incorporated into Hōjō domains.

Ujiyasu, the third Hōjō leader, selected 28 warriors who had loyally served his father and grandfather, with 20 serving as captains, five elders as bannermen, each of whom led a group of a distinct colour, and three other elders whose forces were not so coloured. The five banners constituted the core of the Hōjō armies.

The Hōjō adopted more systematic requests for mobilization around the year 1572. Surviving documents provide a

Through repeated surveys the Hōjō were able to mobilize an increasingly large portion of the population – so that in the late 1580s, their armies consisted of 50,000 men.

The Hōjō surveyed their lands and, based upon the productivity of the land, which was assessed in units of cash, the amount of military service was determined. This system appears to have existed at the time of Ujitsuna, for while writing in his testament of 1541 he describes warriors as possessing a status contingent upon cash revenue. One of clear snapshot of the nature of specific units. To take one example, Okamoto Hachirō Saemon Masahide was responsible for lands worth 59 *kan* of cash, which converted to current currency would constitute 59,000 dollars (29,500 British pounds) and he was required to mobilize in addition to himself, four samurai, and ten *ashigaru* skirmishers. Of these lands, Masahide himself owned lands worth 15 *kan*, while his four samurai possessed one third as much (land worth 5 *kan* each) and the skirmishers owned land worth 2.4 *kan* of cash. Only Masahide himself was mounted, and wore metal armour and a face guard, which was the prerogative of higher-ranking warriors. Two of his *ashigaru* carried flags and six had pikes of 'medium length', being approximately 6m (19ft 8in) long. By comparison, the rival warlord Oda Nobunaga utilized 'long' pikes 9m (29ft 6in) in length. Another document dating from the twelfth day of the third month, shows that Okamoto dispatched three men, presumably from his unit, with one armed with a bamboo pole and two others with pikes. This document criticizes him for not providing any bows, and likewise told him that the armour of his followers needed to be match those of the units to which his men belonged. The Hōjō demanded armour of the right colour, although they did no specify its construction, meaning that perhaps these solders wore paper armour as well.

Miyagi Shirō Hyōe no jō, a substantially more powerful warrior than Okamoto Masahide, also went to battle on horseback, and he mobilized seven other horsemen and 28 foot soldiers in 1572. The ration of horsemen to infantry here was 20 per cent, which was a typical ratio for the Hōjō armies. The total worth of the men in this unit was 284.5 *kan* (284,000 dollars; 142,000 British pounds) with the seven horsemen ranged in worth (expressed in *kan*) from 90 to 13, while the average for these men being slightly under 41 *kan*. Miyagi and his men were mobilized as a rate of one man per 6.9 *kan* of cash, which proved to be much less burdensome a ratio than Okamaoto Masahide had to endure, at 3.93 *kan* per warrior. Whereas Okamoto Masahide did not equip his troops with even bows, however, Miyagi had 17 pikemen, two marksmen with matchlocks, one archer, three bannermen

A scene of the siege of Osaka Castle, 1615, showing units of several different daimyō.

FANS AND BATTLEFIELD CONTROL

Direct command became difficult as army size increased. Command and control remained limited. Commanders held fans (*gunpai uchiwa*) as symbols of their office, which could be used to initiate an attack, although generally these were only used for prayers or other rituals. Takeda Shingen used his fan to save his life by parrying an enemy strike. Incidentally, these objects continue to be used for Sumo wrestling, and became commonly used throughout Japan, a legacy of the intense militarization of society in the sixteenth century. Conversely, 12-ribbed folding fans (*gunsen*) were used in battle as well, often as a means of giving gifts from a commander to a warrior. These fans were also used in other ceremonies, such as those involving the inspection of enemy heads.

Armies became so large, however, that only guiding units in a general direction became possible. The Uesugi, for example, had a set of fans designed to allow for 'war games' to be played, allowing for the movement of larger army units. General commanders were merely responsible for the disposition of their troops, and save for rare cases such as at the battle of Kawanakajima, they did not actually fight at the head of their troops.

A gold-lacquered zunari *helmet with chain mail overlay and decorated with a commander's flag.*

Asakura, a *daimyō* of north-central Japan, similarly emphasized mobilizing large groups of warriors and arming them with pikes. Asakura Toshikage would write in 1480 that even the greatest sword worth 10,000 *kan* can be overcome by 100 pikes each worth 100 *kan*, and therefore, one should purchase large quantities of cheap weapons, rather than a few well made ones.

Eastern Japan witnessed great stability and prosperity under Hōjō rule as they effectively and inexorably occupied much of its land. As a result of prudent policies, the Hōjō constructed a massive castle, something innovative in scale, at the town of Odawara on the coast, and this proved so impregnable that in cases of all-out attack they could simply withdraw and wait out an opposing army. This worked well until 1590, when they miscalculated and upset the warlord Toyotomi Hideyoshi, who besieged them with a massive army drawn from all of Japan and waited them out until they surrendered. The Hōjō

and five others. One final example reveals a warrior with 25 *kan* of cash mobilized three men, one being a horseman, the other having a flag and the third a pike of medium length for a ratio of 8.3 *kan* per person.

Although it would be an exaggeration to describe the Hōjō armies as being a paper force, they did possess inadequacies. They mobilized a force that predominantly consisted of pikemen, allowing them to occupy lands in great numbers. Horsemen were relatively few, and not apparently used to fighting in coordinated groups, and these armies possessed a surprising lack of projectile weapons. To take the three units mentioned above, we can see they comprised 54 men with 10 men on horse and 44 on foot.

Of the latter forces, four were samurai, 24 wielded medium-length pikes, six had only their flags, and a grand total of three men were capable of firing projectiles. Finally seven men appeared in battle with no notable weapons at all. Perhaps they were porters or cooks, or perhaps they were simply men in reserve ready to fight if and when they could capture a weapon.

The basis of these armies consisted of pikemen, and a surprising lack of bows and guns suggests that these forces were primarily intended for land occupation. Many of these men had minimal armour, and the preponderance of pikemen mean that these armies were not particularly mobile. This military organization perhaps explains why the Hōjō only gradually extended their domains through physical occupation, rather than lightning offensives. In this sense, the Hōjō were not unique, for the

Honda Tadatomo (1582–1615), the second son of Honda Tadakatsu (1548–1610) wore similar antlered armour to that of his father, and fought at Sekigahara in 1600. Inebriated during the first siege of Osaka, he fought valiantly against Mōri Katsunaga at the Tennōji gate of Osaka Castle, and was killed. After death, he achieved apotheosis as a god of temperance.

A HŌJŌ UNIT

The forces that Okamoto Masahide mobilized for the Hōjō. Note the preponderance of pike, and the fact that Masahide alone was mounted. Remarkably, he did not provide any bows or guns at all, suggesting that the Hōjō were excellent at mobilizing an army of pikemen who could occupy territory, but were perhaps not the most potent fighting force.

themselves committed suicide, but their domains and samurai otherwise surrendered intact.

Thus, administrative efficiency and innovation in fortifications could not protect armies forever, particularly in light of the rise of firearms and cannon. Resistance to change led to the Hōjō failing to realize that what had been sophisticated in 1500 was largely obsolete by 1590, and they were the last great *daimyō* to fall, thereby bringing an end to the 'Warring States' period. Not all *daimyō* proved as administratively adept as the Hōjō. The Uesugi were, by contrast, less skilled at organizing their domains, but they were better at fighting, and managed to survive, albeit in reduced circumstances, the turmoil of the sixteenth century.

THE UESUGI: THE LIMITS OF CHARISMATIC CONSERVATISM

The Uesugi were an old and illustrious family. Their fortunes were made because one woman, Uesugi Seishi, was the

A nineteenth-century representation of Uesugi Kenshin (1530–78). Born into the Nagoe deputy shugo *family, he was adopted and made heir to the Uesugi. Like his great rival, Takeda Shingen, he took religious orders. Kenshin is the name he had as a lay monk from 1571. He is most famous for his repeated battles at Kawanakajima, and also was a great supporter of the Ashikaga shoguns.*

mother of the first Ashikaga shogun, Takauji, and they became ensconced as the Ashikaga deputy of the east (*kantō kanrei*). This did not prevent them from suffering from chastisement, particularly in the early fifteenth century by Ashikaga Yoshinori, and indeed they never fully recovered from this loss. The alliance of Imagawa and Takeda, to which the Hōjō were added, was directed against the Uesugi, and so effective was this alliance, and steady Hōjō pressure, that the last Uesugi had to abandon eastern Japan in 1552. Uesugi Norimasa (1523–79), who occupied the position of *kantō kanrei*, fled to Echizen in north-central Japan, where a powerful deputy *shugo* family of the Uesugi resided, the Nagao. Here, he adopted Nagao Kagetora as his heir, and this man later became known as Uesugi Kenshin.

The example of the Uesugi reveals that a weakened *daimyō* could protect its name and image by assimilating its most powerful deputy. Kenshin became so devoted to the Uesugi cause that he hardly preserved any Nagao records, but

  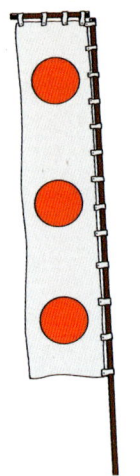 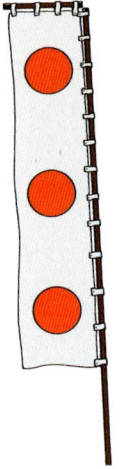

Banners of recognition were inspiring for soldiers and served as points of reference during often confused fighting. From left to right: sashimono *of Makino Tadanari depicting a three-leaved flower (1615);* sashimono *of Tokugawa's retainer Kyogoku Tadatsugu;* nobori *of Sakai Tadatsugu (1527–96); vajra design, Yamamoto Kansuke (16th Century).*

ensured that nearly every scrap of paper and Uesugi heirloom would be preserved, and in this endeavour he was successful, for the Uesugi collection remains unsurpassed.

Kenshin travelled to Kyoto, first in 1554 and then in 1560, in order to be invested with titles from the Ashikaga. He did so, but unlike the Ōuchi he chose not to remain, but instead returned to his domains and then immediately launched an attack on the Hōjō, going so far as to surround their main castle of Odawara and lay siege to it, suggesting perhaps that their defences were more of a paper

COMMANDERS

force than one might imagine. Kenshin was forced to break his siege, as the Takeda threatened him to the rear, but he still managed to travel to nearby Kamakura, where he had himself appointed as the *kantō kanrei*.

UESUGI ORGANIZATION

A document dating from 1575 provides a detailed snapshot of the Uesugi organization, and it reveals both the real strengths and weaknesses and provides an intriguing contrast with the Hōjō. Unlike the Hōjō, Kenshin had no use for creating a clear link between the assessed value in holdings and the number of troops brought into battle. Thirty-nine names are listed in all, revealing that a familial, rather than a regional, basis served to organize troops, and they led a total force of 5514 men. Of these, 3609 were pikemen, 650 were classified as porters (they were literally known as 'open handed' or *teari*), 321 had guns, 368 had banners and 566 were mounted on horseback.

There are some notable trends in the ratios. First, the number of mounted warriors is less than the Latter Hōjō, at slightly under 10 per cent as opposed to 20 per cent (a fact that the mobility of Kenshin, and the stasis of the Hōjō, would suggest otherwise). The Uesugi used fewer bannerman in their forces than the Hōjō, and they mobilized an army that was smaller in size and less extensively organized than the Hōjō forces. Sugiyama Hiroshi suggests that far more lower-ranking warriors were incorporated into the Uesugi army because they relied on fewer mounted warriors and had so many porters, but

A man wearing tōsei gusoku armour reminiscent of that worn by Honda Tadakatsu. The deer horns are important because the festival he is participating in is linked to Kasuga shrine, and deer are thought to be the messenger of this deity.

one cannot so easily draw a comparison between social status and the ability to ride horses. Kenshin, in describing his forces, claimed that it consisted of 8000 men, a not implausible figure because he can document 5500 – perhaps some allies helped join him. Kenshin did not, however, tightly control the regions that he 'owned' and at times his followers rebelled; he might attack them, but ultimately ensured compliance through the taking of hostages.

Although the Uesugi did not organize their province well, evidence suggests that their military organization, rather

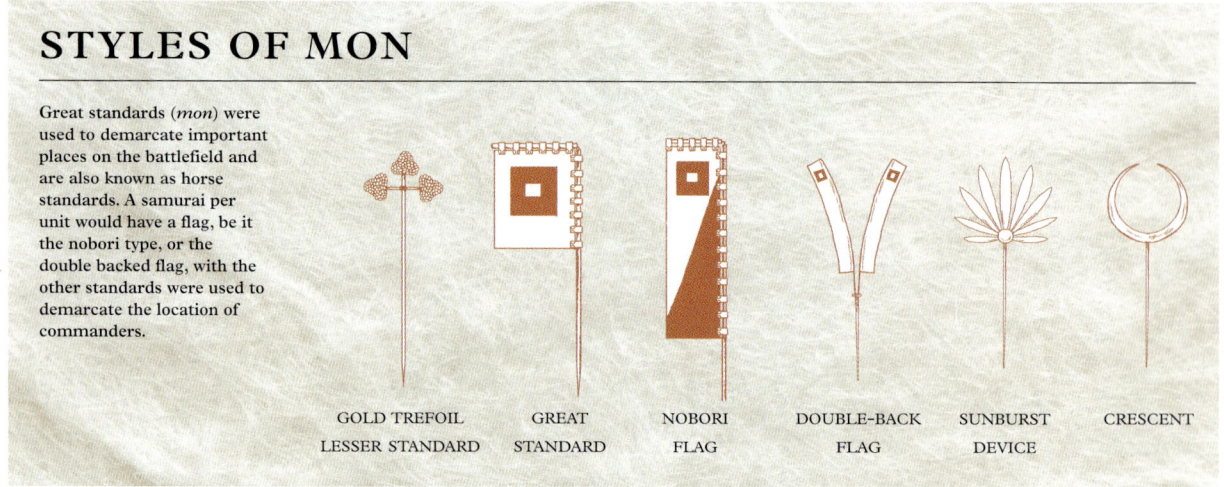

STYLES OF MON

Great standards (*mon*) were used to demarcate important places on the battlefield and are also known as horse standards. A samurai per unit would have a flag, be it the nobori type, or the double backed flag, with the other standards were used to demarcate the location of commanders.

GOLD TREFOIL LESSER STANDARD | GREAT STANDARD | NOBORI FLAG | DOUBLE-BACK FLAG | SUNBURST DEVICE | CRESCENT

than their administrative bureaucracy, proved advanced. A military primer, entitled *The Secrets of Military Organization*, for example, explains how Uesugi Kenshin organized his military encampment as if it were a mobile castle surrounded by a fence, which was itself guarded by numerous warriors, groups of *ashigaru* guarded a rectangular enclosure manned by groups of archers and squads of marksmen. Protected by these concentric barricades, Kenshin resided in a temple that served as the centre of the encampment. His care in fortifying even temporary lodgings would ensure that he, unlike Imagawa Yoshimoto at Okehazama in 1560, would not be negligent and killed in his camp.

Kenshin also proved to be a charismatic leader who saw himself as a protector from the north, and thus linked himself to a Buddhist divinity named Bishamon, who guarded the northern direction, and had his flags emblazoned the character 'Bi' for Bishamon. He also possessed another battle flag, showing the character for 'dragon' drawn in cursive, which would be used to signal a general offensive.

Like the Ōuchi, however, Kenshin seems to have devoted himself to upholding the Ashikaga order. Suits of armour were granted to him from the shoguns, demanding that he come to their aid, and he expended considerable energy in order to uphold old titles of the Uesugi. Unlike the Hōjō, and even more so than the Ōuchi, Kenshin exhausted himself with repeated offensives and travels to Kyoto and Kamakura. Clearly he did not see Japan as a conglomeration of autonomous states, but rather a single entity, ruled nominally by the Ashikaga and in desperate need of a protector.

KAWANAKAJIMA

Five battles fought at the same location of Kawanakajima, at the boundary of the Takeda and Uesugi domains, served to define Kenshin as a charismatic leader. One would think that five battles fought at the same spot represents an exercise in futility and bad generalship, but in fact the intensity of the conflict belies the strategic importance of this region, for it was at the boundary of Shinano and Takeda's Kai Province, where two prominent warriors, Ogasawara Nagatoki (1519–83) and Murakami Yoshikiyo (1501–73) could not stem the Takeda advance and instead called on the Uesugi for protection. The first conflict occurred in the eighth month of 1553 and it was followed by another battle during the seventh month of 1555, another during the eighth month of 1557, the ninth month of 1561, and finally the eighth month of 1564. Kenshin had to fight, for to lose these lands would make his realm

Uesugi Kenshin leads a charge against the forces of Takeda Shingen at the fourth battle of Kawanakajima. Kenshin has a shaved head and is located at the centre of the image.

vulnerable to the Takeda. Furthermore, the mountain castle at Kasugayama was Kenshin's staging point for his attacks in the *kantō*, eastern Japan, and so to lose this area would mean that Kenshin could no longer attack the Hōjō in the east and assert his position as 'protector of the east'. Of all the conflicts, the battle in the ninth month of 1561 proved to be the most intense, and probably the most famous battle of the century. That being said, surprisingly few sources survive that recount this battle, though it appears to have been a battle of manoeuvre. Kenshin's force of 18,000, or so the chronicles claim (some three times more than can be verified), first occupied the mountain of Saijosan, which he had never occupied during the previous encounters, and constituted his most aggressive manoeuvre to date. Shingen approached with a larger army of 20,000, which is perhaps an exaggeration, and advanced beyond Saijosan, occupied the nearby Cha'usu mountain and then advanced to Kenshin's rear, readying an attack on Saijosan.

There the armies watched each other for four days, before Shingen suddenly advanced beyond Saijosan, to the north, concentrating his forces at Kaizu Castle in a move known as a 'woodpecker feint' – for just as a woodpecker would poke its beak in a hole so as to force out an insect from a tree, so too was this manoeuvre intended to cut off Uesugi Kenshin from his base in the north entirely, and force him to move. Shingen concentrated his main forces at Kaizu but sent a smaller force of 8000 men to the south of Saijosan, so as to surround Kenshin.

Kenshin did move, however, and in a dramatic night manoeuvre he passed by where Shingen's forces had been encamped before they moved to Kaizu and turned to the north, so that he could face Takeda Shingen's army face on, minus the estimated 8000-man force that now, pointlessly, guarded the abandoned Saijosan Castle. Shingen's main army of approximately 12,000 had no choice but to fight the superior force.

OPENING MOVES

Little is known about the battle itself, for surprisingly few reliable sources survive. It went, perhaps understandably, very badly for the Takeda initially, and Uesugi forces were able to enter the main Takeda encampment. Kenshin, writing after the battle to the noble Konoe Sakihisa, would himself state that he had earned a name for himself throughout the realm by exchanging blows with Shingen; in this document he likewise argued that he had killed or captured 8000 of the

DATE MASUMUNE

Known as the 'One Eyed Dragon' (*dokuganryū*), Masamune was one of the most charismatic *daimyō* of the late sixteenth century. His helmet has a distinctive crescent moon, and he wears very large shin guards (*ōtate-age sune-ate*) that were popular from the late fourteenth through the sixteenth century, and his cuirass is made of five hinged sections (*gomaidō*). Masamune founded the town of Sendai, and was so powerful that the Tokugawa did not permit him to built a castle with a prominent keep (*tenshukaku*).

Takeda, which seems to be an exaggeration. Kenshin wrote such lavish praise in his documents written after the battle, and the ink used is of an unusual hue, leading many to think that they are documents 'stained in blood'.

These records, written with feverish emotion by Kenshin referred to a 'great victory with several thousand of the enemy taken' in letters to Yasuda Nagahisa and Irobe Katsunaga. And, according to the *Myōhōjiki*, Shingen's brother Nobushige was killed in battle, while the Uesugi themselves also apparently suffered heavy casualties. Shingen himself claimed that they suffered 3000 casualties in a letter written at the end of the tenth month of that year.

We know that the battle was fought with intensity, resulted in high casualties and that fighting occurred inside the main encampment where Takeda Shingen was located. All major accounts of this battle written after these events,

Saitō Toshimitsu, a general in the army of Akechi Mitsuhide, reconnoitres along a river leading to the capital, Kyoto, in 1582. Mitsuhide's forces overwhelmed those of Oda Nobunaga, but were in turn destroyed two weeks later.

from both the Takeda and Uesugi side, state that this encounter occurred. It is conceivable that, sensing the chance of killing his rival, Kenshin himself fought with Takeda Shingen.

It has not been proved that the two actually exchanged blows, but it became widely thought that they did, and this in turn enhanced the cult of personality around these two charismatic generals. Indeed, personality gave an edge to their conflict, for it seems that Kenshin actively despised Takeda Shingen.

He wrote in 1564, during his fifth encounter with Shingen, a prayer to a shrine recounting Shingen's evil deeds, including the destruction of temples and shrines in Shinano and other regions, and the fact that Shingen had banished his own father and forced him to become a beggar. This prayer, and later accounts of Shingen, characterize him as a secular figure, but such perspectives are misleading. Shingen was the highest-ranking Tendai Buddhist priest in his home province, and he was visibly upset when Oda Nobunaga burned the Tendai headquarters of Enryakuji (Mt Hiei) in 1571.

The Kawanakajima battle has been immortalized on numerous screens, one of the most impressive of which remains with the Uesugi to this day. This screen proves illuminating, because it reveals primarily that the battle was fought in close quarters by pikemen, while archers peppered the enemy with arrows. We also know, from other sources, that the Uesugi had been exposed to guns, for in 1560, they received a detailed explanation of how to mix gunpowder from a messenger of the Ashikaga shogun, Yoshiteru, which reveals that their interest in the Ashikaga, and ties to the capital, could have some very real advantages.

CONCLUSION

To conclude, our survey of commanders reveals a variety of approaches and attitudes. The Ōuchi modelled their behaviour and organization on the Ashikaga. Although initially powerful, they weakened over the course of the sixteenth century, and were ultimately destroyed by a rebellion of their deputy in 1551.

By contrast, the Hōjō provided a model for later samurai attitudes, and also institutional development, and their armies were large, but poorly equipped and slow. Unlike the Hōjō, the Uesugi did not possess as intricate a system of organization, but their armies proved more mobile, and their charisma and astute alliances helped them to survive.

In order to better understand the Uesugi's luck, and the Hōjō's lack of it, it is necessary to turn to the dissemination and role of the guns. For those daimyō who used them most effectively, particularly the Oda and to a lesser degree the Uesugi, met with success, while others such as the Takeda and, to a lesser degree the Hōjō, did not. An exploration of the gun reveals both the importance of a new technology and its limitations in transforming a society.

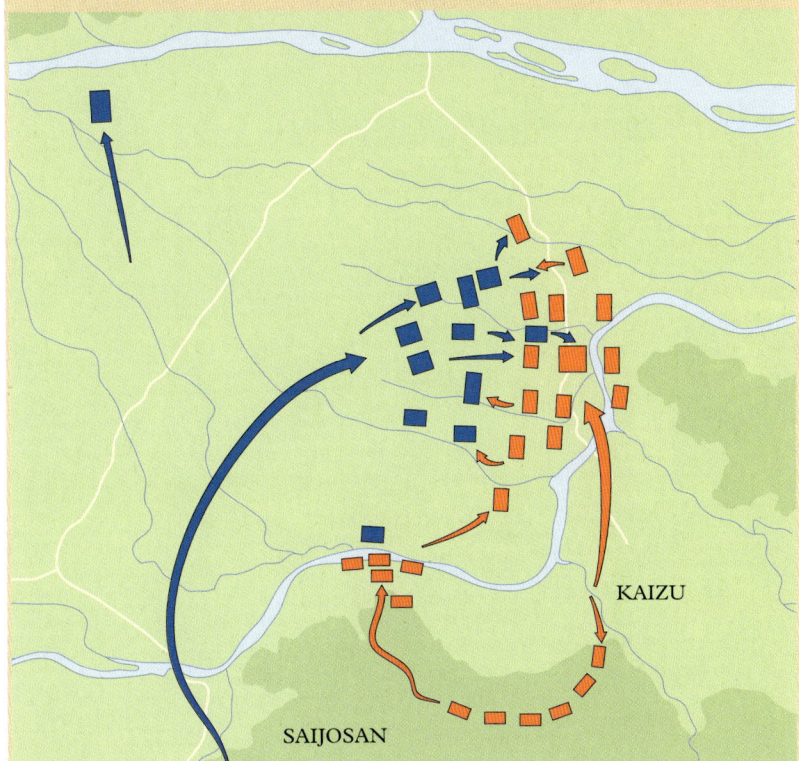

FOURTH BATTLE OF KAWANAKAJIMA

Kawanakajima was a strategic area that the Uesugi had to control in order to be able to march to the capital, and the site of five battles between the Uesugi and the Takeda. Some of these 'battles' were mere skirmishes, but the fourth battle of Kawanakajima nearly caused the Takeda's annihilation. Uesugi Kenshin (blue forces) surprised Takeda Shingen (red forces) by marching at night from his initial position at Saijosan Mountain to the north, and forced him to attack. Shingen had divided his forces, and the contingent he dispatched to the south, to guard against Kenshin at Saijosan, was so removed from the battle that it could only join the battle very late. Its belated arrival rescued Shingen from decisive defeat and possible death. Although the Takeda suffered heavy casualties, their forces inflicted substantial casualties on the Uesugi as well.

Firearms

Shortly after a chance landing in 1543 on Tanegashima, a small island 71km (44 miles) southeast of Kyushu, Portuguese merchants parted with three of their firearms (TEPPŌ). Two of these went to a local lord, named Tokitada, while a priest affiliated with Negoroji Temple purchased the third.

Samurai in traditional armour prepare to fire their guns, signalling the start of the Musha gyōretsu *festival at Odawara Castle.*

These objects were, according to Nanpō Bunshi's *Record of the Musket*, 'about one metre [3ft 3in] long, straight on the outside and hollow inside, made of a heavy substance' with a hole on its side, where fire was applied. The tube was filled with 'miraculous powder' and a 'lead pellet' and when the gun was fired 'the pellet … hit the target … [and] the explosion seemed like lightning and sounded like rolling thunder.'

NEW TOOLS OF WAR

Record of the Musket was written for Tanegashima Hisatoki (1568–1611), the lord of this island, some 60 years after the events described. Bunshi paints a picture of the inhabitants of Tanegashima not even knowing what to call it, and that they settled on the word *teppō*, for gun. The newness of the gun to Japan also appears in an account of a Portuguese merchant, Mendes Pinto, who claimed that 'the Japanese had never seen firearms like [the musket] before.' Here Pinto is perhaps more revealing, in that the Japanese of Tanegashima were well aware of primitive firearms, which had been used for nearly 90 years on the Ryūkyū islands to the south of Tanegashima.

Hyperbole detracts from Bunshi's account, for he invariably describes marksmen hitting with 100 out of 100 shots in target practice. Pinto's account is dogged by inaccuracies, such as his description of the payment by the lord of Tanegashima of a thousand silver *taels*, a Chinese currency that was not used in Japan. Nevertheless, amid these errors, both accounts reveal a fascination with the Portuguese gunpowder and an equalled interest in the guns per se. Mendes Pinto wrote that 'none of them knew the secret of the gunpowder and could not understand how it worked, [so the Japanese] attributed it to some sort of witchcraft.' Likewise, the *daimyō* of Tanegashima spent a princely sum to the merchants for the weapon and, according to Pinto, 'begged him to teach him how

A Japanese depiction of a Portuguese ship. The Portuguese, known as 'Southern Barbarians' because they first came to the island of Tanegashima from the south, impressed the Japanese with their guns and impressive ships. Not all artists, however, were capable of depicting their vessels accurately, as this screen reveals.

JAPANESE HARQUEBUS AND SIDEARMS

This photo reveals how small swords and some nineteenth-century firearms compare in size. Compare the *tantō* dagger, two *wakizashi* and a *katana* with the following nineteenth century arquebuses (*bajōzutsu* – literally, 'horseback barrel'), designed for firing on horseback, which continued to use an old fashioned firing mechanism. The shorter version was called a *tanzutsu*, literally, 'short barrel'. These handheld harquebuses could be easily concealed in a samurai's robes. They fired bullets ranging in weight from 11 to 30g (0.3–1oz) as far as 30m (98ft), but could only kill or wound at a maximum distance of five metres (16ft).

to make the powder, for without it, the musket was just a useless scrap of metal.'

These guns would become widely disseminated in Japan over the course of the sixteenth century. Writing with his penchant for exaggeration, Pinto stated that 'when we left the island some five and a half months later, there were already more than 600 of them around.' Pinto estimated that by 1556, 30,000 guns existed in the northern Kyushu province of Bungo and that 300,000 could be found in all of Japan as 'the land became so flooded with them that today there is not a village or hamlet, no matter how small, where they do not produce a hundred or more, and in the important cities and towns they speak of them in nothing less than the thousands.' Pinto exaggerates, but as we shall see, guns spread rapidly throughout Japan after the first Portuguese encounter in 1543.

Historians have long perceived this contact to be epochal. Standard narratives of Japan's sixteenth-century history portray regional lords or *daimyō* as being the most cognizant of the power of these new weapons, and most able to use them effectively. Oda Nobunaga, the first of the 'three unifiers' of Japan, has been characterized as a military genius whose concentrated use of firepower allowed him to 'revolutionize' warfare, crush his most potent rival, the Takeda of Kai province, and consolidate power from 1570 until his untimely assassination in 1582.

As we shall see, the introduction of the harquebus proved significant, for these more powerful guns tripled the distance at which armoured men could be killed on the battlefield. These new guns supplanted earlier, more primitive firearms, which had existed in Japan for over a century, as well as bows. Harquebuses did not, however, lead to the supremacy of infantry, for this

process had occurred well before the arrival of the Portuguese.

Daimyō from central and western Japan gained an advantage over their eastern rivals with the introduction of the gun, for they were more readily able to purchase such weapons, as well as import the ingredients of gunpowder. *Daimyō* and temples from central Japan proved most effective at organizing gunners, and achieved superiority on the battlefield during the course of the 1570s.

We shall first explore the nature of the early guns, as well as the composition of the gunpowder used in them, before analyzing the Portuguese weapons and a more effective recipe for mixing gunpowder. Then we shall explore improvements in military organization before recounting the battle of Nagashino, which was fought in 1575 between the forces of the Oda and the Takeda, and resulted in a crushing Takeda defeat.

THE EARLIEST FIREARMS

A bewildering variety of terms describe early firearms. *The Record of the Musket* is wrong in stating that guns were so new that they did not have a name. In fact, guns had been known for some time, and were identified according to a variety of names. Explosive shells, also known as *teppō*, had been known in Japan since the thirteenth century. Mongol forces invading Japan in 1274 and 1281 used explosives to scare Japanese defenders. These objects, which archeologists uncovered in 2001, consisted of hollow metal shells filled with a material presumed to be gunpowder. Little is known about these weapons or their efficacy, but a few chronicles describe how they surprised the Japanese defenders with their loud report.

What has been thought to be the oldest visual depiction of an exploding projectile, found in the thirteenth-

OPPOSITE: *Portuguese and Japanese traders, as depicted in the sixteenth-century* Nanban byōbu *screens. Not all merchants were missionaries, but trading and proselytizing were closely linked.*

Men firing harquebuses from the Budō Geijiutsu hiden zue *(1855). Keeping the powder and matchlock dry proved difficult. Here men with straw raincoats try to fire their guns in inclement weather.*

century *Scrolls of the Mongol Invasions of Japan*, was in fact added to the scrolls during the eighteenth century. It remains difficult to ascertain if these weapons were simply meant merely to surprise others with their sound, or were intend to explode and wound others. As the Mongol forces also relied on catapults, and used far more solid stone projectiles than hollow metal ones, it would seem that the early *teppō* were of limited military significance.

CHINESE GUNS

References exist in Chinese accounts to a primitive gun known as a 'fire dragon spear' (*hiryūsō*) being created in 1355, and the oldest surviving specimens are thought to date from the 1370s. Some black and white illustrations of Ming

147

CHINESE FIREARMS

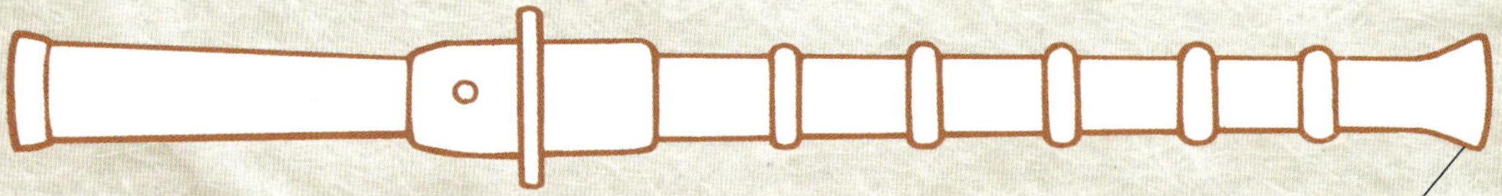

ABOVE: Sixteenth-century 'Silk' gun. The name of this Chinese gun refers to the way that the iron barrel was covered with raffia and then wrapped in silk, making for an extremely attractive light artillery piece.

LOWER: Early triple-barrelled Chinese gunpowder weapon. Accuracy from this firearm must have been extremely poor – note the stock, which would have been clamped under the armpit. Each barrel has a separate vent hole for firing.

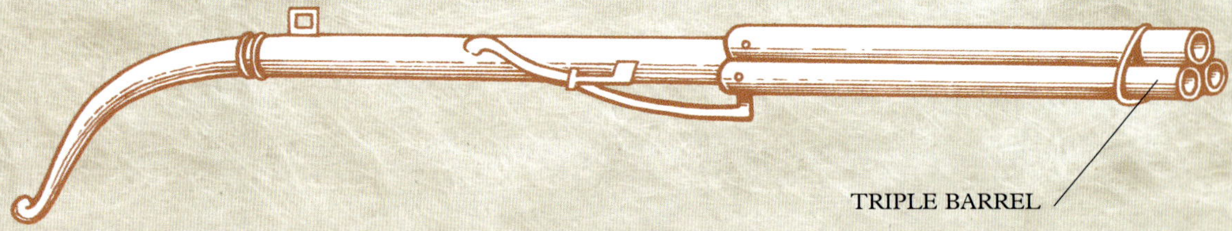

BARREL END

TRIPLE BARREL

A fire spear, one of the primitive firearms that may have been used during the wars of 1467–77.

army tactics, dating from 1551, also reveal Chinese soldiers firing these three-barrelled weapons.

These three-barrelled weapons were transmitted from the Ryūkyū Kingdom, which comprises modern Okinawa, to Japan in 1466, some 80 years before the Portuguese landed at Tanegashima. A fifteenth-century diary records how an official from the Ryūkyū Kingdom surprised many bystanders in Kyoto with the report of his firearm on 28 July 1466, just before the onset of the Ōnin War. Unsurprisingly, these weapons were used during this conflict, for Unsei Daigoku described how a 'flying projectile fire spear' (*hihō hisō*) was discharged from a besieged tower.

The 'fire spear' appears to have represented a different weapon than the three-barrelled projectiles, and the research of Joseph Needham, an erudite and prolific scholar of Chinese technology, reveals that this object could have been a bamboo tube that emitted fire and arrows. In Chinese sources, these three-barrelled guns were sometimes referred to as 'three-eyed lances' and so it is difficult to know if one or several types of primitive firearms arrived in Japan at this time.

Archeological evidence from Akenajō and Katsurenjō, two Okinawan castles, reveals that guns were used prior to the mid fifteenth century. Defenders supplemented the weakest point in Akenajō's defences with a portal especially designed for use by snipers, placed low in its stone walls. Portals in the walls of two castles, Akenajō and Nakagusukujō, are located within 45cm (18in) of the ground, and are thus too low to allow for arrows to be fired. Furthermore, archaeologists have uncovered a total of 11 stone bullets, while one made from fired earth and another of iron have been uncovered

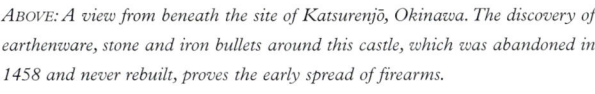

ABOVE: A view from beneath the site of Katsurenjō, Okinawa. The discovery of earthenware, stone and iron bullets around this castle, which was abandoned in 1458 and never rebuilt, proves the early spread of firearms.

RIGHT: The walls of the castle, as seen from below.

within the battlements of Katsurenjō, which was destroyed in 1458 and never rebuilt. At other sites, bullets made from limestone, coral and sandstone, as well as copper and iron were also discovered.

'FIRE ARROWS'

Primitive firearms composed of three metal tubes welded together, known as 'fire arrows' (*hiya*) were widely disseminated in Europe and Asia throughout the late fourteenth and early fifteenth centuries. Remarkably, some of these fifteenth-century *hiya* were fired as late as the early twentieth century, and they were reportedly capable of blasting projectiles for 200m (656ft). These projectiles seem to have been used primarily to break down fortified walls.

TŌSEI GUSOKU ARMOUR (1560)

Tōsei gusoku armour, late sixteenth century. This suit is attributed to belonging to the Christian *daimyō* Naitō Tadatoshi of Tanba (1550?–1626) who fought in Korea in the 1590s and at the battle of Sekigahara in 1600. He converted to Christianity, adopting the name Jo-an (a homonym for John) and was banished to the Philippines in 1614 because of his conversion. His armour eventually found its way to Madrid, where it was catalogued as Moorish armour, but was recognized as being Japanese armour in the ninteenth century, when it was moved to the Tower of London's Royal Armouries. The cuirass of *tōsei gusoku* armour, which became popular in the latter half of the sixteenth century, was divided into two (*nimaidō*) or five sections (*gomaidō*), and has much simplified lacing as compared to earlier versions.

ZUNARI HELMET

HŌ-ATE

OKISODE

NAITŌ FAMILY CREST

SUSO-ITA

ROCK INJURIES

Most primitive firearms discharged rounded stones, and intriguingly data drawn from sixteenth-century military petitions reveals a sudden upswing in rock wounds, particularly in western Japan, which suggests the dissemination of such weapons. Even though only a handful of cases where soldiers were wounded by rocks can be documented in the fourteenth century, rocks injured 82 men during the years 1524–52, with over half (44) occurring during the seventh month of 1552.

Archeological excavations of surviving Okinawan castles reveal that numerous projectiles peppered the walls of such fortifications, the largest being the size of a softball.

The first documentary evidence describing wounds caused by firearms appears in Amano Okisada's battle report (*kassen chūmon*) of 27 November 1527, where one man is listed as being 'shot [and] wounded' in the right foot by an unidentified object. The verb 'shot' (*iru*) referred to bullet wounds, for similar language appears in Okisada's letters, dating from 1569, which explicitly mention guns (*teppō*). Furthermore, Okisada described wounds caused by arrows differently in a document he submitted on 13 May 1527.

Guns were more commonly used in western Japan. Eastern Japan, being farther from the Asian trade routes, witnessed fewer guns, although according to the *Five Generation History of the Hōjō* (*Hōjō Godaiki*), Hōjō Sōun does appear to have received a *teppō* originating in China, from a monk in 1510. According to histories of the Takeda (*Kōyō gunkan*), the Takeda likewise received such a weapon in 1525. The Takeda and Hōjō weapons were probably the primitive three-barrelled guns that can be documented as existing in the capital in 1466. These weapons could impress with their sound, but they were not accurate, and did not even possess the maximum range of a longbow, for their bullets travelled 200m (218 yards), whereas longbows could be fired, admittedly inaccurately, as far as 350–400m (382–437 yards).

The limited range stemmed from several factors. The barrels were not rifled and most early bullets were made of hardened earth or rock and were not of a uniform size. This poor ball to bore fit allowed for explosive gas to disspate.

Europeans dramatically improved firearms over the course of the sixteenth century and, befitting such a period of innovation, myriads of different types and terms existed – harquebuses, archibuso, arcabouza and in English hackbuss, hagbutt and hakbutt, as well as the term firelock, matchlock or musket. Musket, a common term for these weapons, arose in Italy in the 1530s, and was used to describe heavy weapons that relied upon hemp rope for firing. There were subsequent innovations in the firing mechanism, such as the wheellock, which

Early European firearms being used in a siege during the Hussite wars (1420–34). Although they made a lot of noise and undoubtedly frightened their intended targets, 'handgonnes' were ineffective weapons.

WARRIOR MONK (1570)

The monks of Negoroji fielded a formidable fighting force of 300 gunners in the 1570s. They gained access to Portuguese weapons quite early, and used them effectively in battle, defeating even Oda Nobunaga's brother in battle. Toyotomi Hideyoshi mobilized a large army and incinerated most of the main temple complex in 1585. The sixteenth-century armour depicted here tended to be constructed of continuous rows of scales (*iyozane dō*), with less weaving than earlier styles.

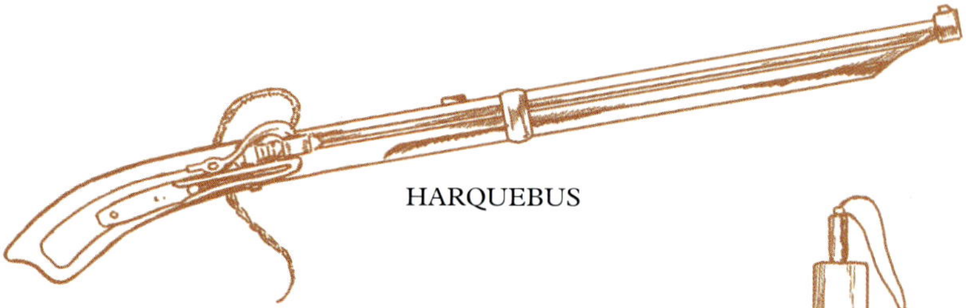

HARQUEBUS

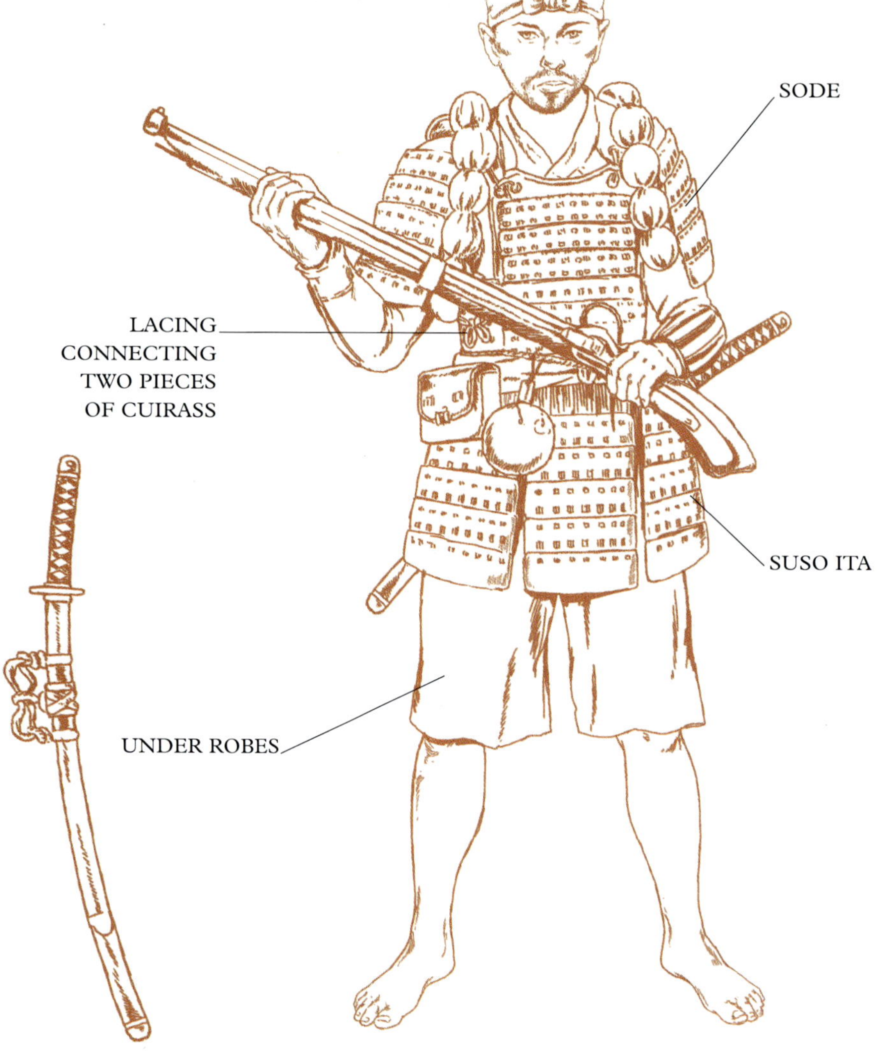

POWDER FLASK

SODE

FUSE REEL

LACING CONNECTING TWO PIECES OF CUIRASS

SUSO ITA

BULLET POUCH

UNDER ROBES

MUZZLE POWDER FLASK

KATANA

relied upon sparks for firing, but these delicate weapons were not widely adopted. Gradually, later types of muskets had a less exaggerated curved butt. One, called the caliver, was developed in 1569. Lighter and with less penetrating power than harquebuses, they do not seem to have been used in Japan.

Harquebuses were introduced to Japan by the Portuguese and became the favoured weapon. They were common in Europe in the fifteenth century, and by the sixteenth century they developed a distinctive curved butt, and also had a particular firing mechanism which used an s-shaped piece of metal called a serpentine, which contained a match, and would, once triggered, light priming powder that was sprinkled beforehand into a pan at the side of the barrel. This flash penetrated a small hole in the barrel and lit the main charge. Serpentines had constituted the firing mechanism of crossbows in Europe, and the adoption of this trigger mechanism improved early guns remarkably.

The distinctive curved butt of harquebuses was designed to raise the barrel of the gun to eye-level, and so allow for more accurate aiming. Over time, however, a square butt would become more common, for it allowed the recoil of a rifle to be absorbed by the marksman's shoulder. The Japanese preferred the curved harquebuses, however, and strove for accuracy in firing rather than increasing the rate of fire. Still, as considerable overlap existed in terminology, we shall also use the words musket and firearm interchangeably, with harquebus to describe these heavy Japanese weapons with a curved butt and a firing mechanism relying on hemp rope and a curved serpentine.

THE SPREAD OF THE PORTUGUESE HARQUEBUS

Portuguese harquebuses were spread by the Ashikaga shoguns, who received them from the Tanegashima *daimyō* and gave them in turn to their supporters. The Ashikaga would also pass gunpowder

recipes on to favoured *daimyō* as well.

Writing sometime before 1549, Ashikaga Yoshiharu thanked Hosokawa Harumoto, saying in a document that survives in the Honnōji records that 'a gun [*teppō*] arrived from Tanegashima. How joyous. I will write a letter to [the] Tanegashima [*daimyō*]. Please send it to him.' And these guns seem to have influenced battles the following year (1550) in the capital, for according to Yamashina Tokitsugu's diary, he witnessed a skirmisher, Miyoshi Yūsuke, who was shot by a gun by a member of

*Ōtomo Yoshishige (1530-87), scion of an old family of constables (*shugo*) of Bungo Province, became a Buddhist lay priest in 1562, taking the name Sōrin. In 1578, he was baptized and took the name Ōtomo Francisco. Having converted to Christianity, he received several Portuguese cannon, but he was nevertheless badly defeated by his rivals, the Shimazu of Southern Kyushu.*

Hosokawa Harumoto's forces and perished. Other accounts, found in the *Nochikagami*, reveal that Harumoto built a castle at Nyoigadake, one of Kyoto's south-eastern hills (and site of the 30

June 1336 battles). This structure consisted of two walls of earthworks, and in between them 'stones were had been added to the walls to protect against *teppō.*' This account corresponds well with archeological evidence, for surviving abandoned castles such as Shōrakuji reveal that stones began to be added late in the 1540s.

Ōtomo Yoshishige (1530–87) is documented in his family's house records as giving harquebuses to Ashikaga Yoshiteru in 1553, 1556 and 1557. Ōdachi Harumitsu, one of Yoshiteru's administrators, praised their quality on 19 January 1554, when he wrote that 'we have many weapons, but this one is unsurpassed' and that 'it will be part of our secret repository' of weapons. Maintaining such a repository did not prevent Yoshiteru from distributing these firearms to others, such as Yokose Utanosuke Narishige of Kōzuke Province in 1553, or, for that matter, disseminating gunpowder recipes to stalwart *daimyō* supporters such as Uesugi Kenshin.

Daimyō without such connections to the Ashikaga had to rely on a variety of stratagems to secure these weapons. The Matsura, a small *daimyō* of northern Kyushu, had two of their samurai convert to Christianity so as to secure the trust of the Portuguese and learn how to manufacture gunpowder. They also received a cannon and three guns from the Portuguese. Francis Xavier, a Jesuit missionary, also describes how he gave three muskets (*suisekijū*) to Ōuchi Yoshitaka in 1551. These had been intended as gifts for the sovereign of Japan, but Xavier improvised after he was prohibited from visiting the capital.

THE MANUFACTURE OF FIREARMS

So sophisticated were Japan's smiths that almost immediately they were able to craft guns that were the equal of those being produced in Europe. Serendipity had its role as well, for Tanegashima, site of the first Portuguese landing, is rich in iron as its black sandy beaches are composed of iron ore. The Tanegashima

Modern Sakai. The port represented a Japanese manufacturing hub in the sixteenth century, and is now incorporated into Osaka.

smiths were well placed to begin manufacturing these weapons, and, as a measure of their success, these weapons came to be known as 'Tanegashima' throughout Japan. Tanegashima was not the only site where muskets came to be manufactured, for they were also crafted in northern Kyushu as well.

Having procured a Portuguese harquebus in 1543, the priests of Negoroji and their affiliated metalworkers soon established a forge of gunsmiths, and produced enough weapons to form a force of 300 marksmen (*teppō shū*) in the 1570s. Negoroji seems to have been one of the most significant centres of gun production and gun use throughout the

St Francis Xavier (1506–51) and his entourage, detail of the right-hand section of a folding screen depicting the arrival of the Portuguese in Japan, Kano School (lacquer finish).

JAPANESE RIFLING AND HARQUEBUSES

Japanese smiths rapidly mastered the most difficult techniques of rifling the barrels of guns. This proved significant, for guns fired more effectively when their barrels were rifled. Smoothbored barrels allowed for bits of gunpowder, lead (or earth depending on the nature of the bullet) to build up, causing guns to misfire. Rifling is thought to have originated in Leipzig in 1498, but no reliable evidence exists for such weapons prior to 1544, with the oldest surviving examples dating from between 1550 and 1560. The existence of the Portuguese Tanegashima harquebus is significant, and under appreciated by European specialists, because it represents the oldest surviving example of a rifled firearm. This Portuguese weapon is capable of firing a 17mm (.67in) round, and had a barrel that was 100cm (3ft 3in) in length. Japanese harquebuses exist in great abundance, and although some were shorter, at 87cm (2ft 10in), longer weapons came to be preferred, ranging in length from 120–140 cm (3ft 11in– 4ft 7in), while one of the largest guns, created in the early seventeenth century, was 300cm (9ft 10in) long and weighed 135.8kg (300lb).

TEPPŌ (SEVENTEENTH CENTURY)

MATCHLOCK MUSKET (EIGHTEENTH CENTURY)

FIRING AN HARQUEBUS

STAGES OF FIRING

It required time to fire an harquebus. Some gunners may have been stationed in groups of three, although they were most often interspersed with archers, who could fire faster, and pikemen, who could offer them protection. Some men were required to carry bullets and a supply of gunpowder. Here, a group is shown in the various stages of supplying, loading and firing a gun.

PREPARING THE HARQUEBUS

In firing, first gunpowder was placed in the barrel (1), and then a bullet added (2), which was rammed in the barrel (3). Powder was then added to the pan (4) and finally the matchlock was fired (5).

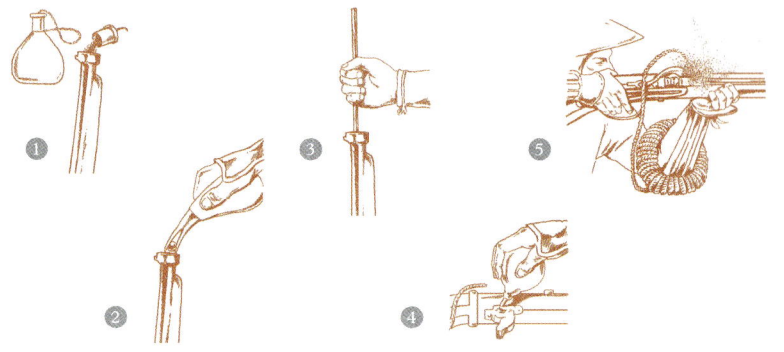

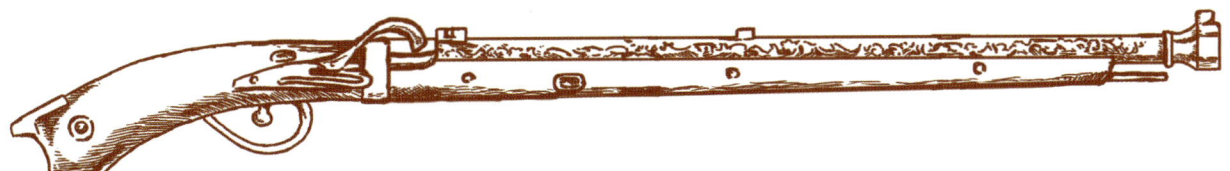

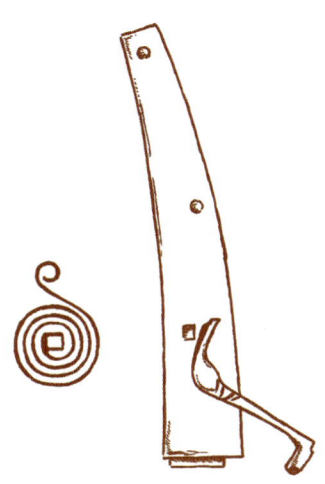

Matchlock harquebus with firing mechanism (external brass spring) – early seventeenth century. These firing mechanisms closely resemble earlier firing mechanisms for crossbows.

1570s. The town of Sakai, located to the south of the capital and northwest of Negoroji, likewise became a hub of manufacturing. The old capital of Kyoto did not have any gun manufacturers. This is evident in a letter from Ashikaga Yoshiteru to Yokose Narishige in 1553. Yoshiteru wrote that he had summoned a smith from 'the south' [of the capital where Negoroji is located] to manufacture a high-quality gun.

It is not known when Japanese craftsmen were able finally to manufacture guns on their own, but they were already successful smiths, being able to manufacture excellent swords for centuries. They easily could manufacture barrels by rolling steel, and one group of smiths, the Kunitomo, managed to devise a process where a screw could be slowly twisted into a red-hot barrel to create rifling. A plug would be screwed into the end of the gun and the outside of the barrel shaped into an octagon. After that a smith would then add the firing mechanism. The continued existence of the Kunitomo smiths allows us to know much about how they manufactured their weapons. Rifling remained an exceedingly difficult task for European smiths to accomplish in the early sixteenth century. Although the English used what was called a 'skowering stick' to ram in the barrel and rifle it, the idea of screwing the barrel does not seem to have become common in England until 1635, when a patent to 'rifle, cut out and screw barrels' was granted. By contrast, the Kunitomo smiths appear to have rifled their weapons by the turn of the seventeenth century.

GUNPOWDER

For guns to be effective, a stable recipe of gunpowder proved as important as a well crafted barrel and firing mechanism. The composition of gunpowder determines the strength of its explosion, and the distance that a bullet can be fired. Potassium nitrate, or saltpetre as it was known, provides the explosive potential of this powder, while sulphur lowers the ignition temperature and also increases the speed of combustion; carbon binds the other two ingredients together. John Bate, whose *Mysteries of Nature and Art: in foure severall parts, the second of fire works* was published in 1635, characterized the ingredients as follows: 'The Saltpetre is the Soule, the Sulphur the Life, and the Coales the Body of it.'

Older recipes of gunpowder contain lower ratios of saltpetre, which meant that they exploded more weakly. Gunpowder originated in China, but the early tenth-century recipes only contained 50 to 55 per cent saltpetre. Gradually, a higher percentage of saltpetre came to be used, with the optimum ratio perfected by the Europeans in the fifteenth century. It is a sign of the importance of this mixture that variation remains concerning the proportions of ingredients. Joseph Needham portrays the ratio as consisting of 75 parts saltpetre to 13 parts sulphur and 12 parts coal. Some variation exists, for the nineteenth-century German general staff, by contrast, felt that the ideal ratio was 74:10:16 or 74:12:13. Fifteenth-century European recipes approach this optimal ratio, although some contain significantly less saltpetre in order to mitigate the strength of the explosion.

A seventeenth-century writer quoted by Joseph Needham writes:

> *The whole Secret of the Art [of making gunpowder] consists in the proportion of the Materials and the exact mixture of them, so that in every the least part of Powder may be found all the Materials in their just proportion; then the Corning or making of it into Grains; and lastly the Drying and Dusting of it.*

The author continues 'you make good *Powder* with any of the proportions above mention'd; but the more [salt] Peter you allow it, it will still be the better, till you come to observe eight Parts', which would refer to between 72 and 78 per cent saltpetre. This ratio would prove most powerful, and one which was used in Japan. Early marksmen underestimated the strength of this

WEAPONS AND FIGHTING TECHNIQUES OF THE SAMURAI WARRIOR

This Yoshitoshi print from 1866 illustrates an episode in Tokugawa Iemochi's ill-fated campaign against the western provinces of Chōshū. Sato Masakiyo does not appear in the fourteenth-century Taiheiki, *but this print was labelled such so as to evade Tokugawa censorship. The violent gunpowder explosion, a common danger, is represented here most dramatically.*

gunpowder, and Mendes Pinto recounts how the son of a *daimyō* nearly severed the thumb of his right hand after the musket that he was firing exploded.

One can get a close approximation of the gunpowder recipe used at Tanegashima, for these weapons and knowledge of gunpowder soon spread to the Ashikaga shogun who, according to the Tanegashima genealogy, had heard that they had 'directly received it from the Southern Barbarians, and that the mixture of gunpowder proved most magnificent'. This recipe became in great demand, and the court noble Konoe Tane'ie wrote to Tanegashima Tokiaki

that 'you have received peerless gunpowder from the Southern Barbarians. Please notify the Shimazu of this recipe, and have them notify the Ashikaga *bakufu* of its composition.'

Ashikaga Yoshiteru had, as we have already seen, access to 'numerous' guns, and furthermore he requested saltpetre from Ishiyama Temple in 1552. The Ashikaga soon passed on knowledge of this gunpowder to their favoured supporters. On 29 June 1559, Ōdachi Harumitsu, acting on behalf of the shogun, dispatched a copy of the Ōtomo Yoshishige's gunpowder recipe to the ailing Uesugi Kenshin. This document reveals an extremely potent mixture with the ratio of saltpetre to sulphur and coal being provided twice, at 80:12:8 and 77:13:10, with the latter ratio in particular approaching the optimal. It seems likely that the latter rate was used, for it required less saltpetre, which Japan did not possess in abundance. Japan,

Upper Ishiyama Temple, which was a source of saltpetre for the Ashikaga Yoshiteru in 1552.

SOURCING POTASSIUM NITRATE

Potassium nitrate forms when high temperatures and high humidity allow for organic matter, especially excrement, to decompose. Some regions have this naturally occurring – in caves in Laos, for example, bat guano provides a ready source of this material. Europe proved less conducive for the production of saltpetre, which it extracted from toilets, where the urine of wine drinkers was favoured over all because of elevated nitrate content. And in Japan too, Mōri Motonari, writing to his followers in the spring and early summer of 1554, would state that the earth where old stables had been located was ideal. As the Uesugi had less ready access to these trade networks, they probably used less saltpetre, which in effect, meant that their gunpowder was combined at an ideal rate.

A gunpowder flask, depicting a 'Southern Barbarian' and epitomizing the association of Europeans with gunpowder weaponry.

unlike China, did not have large stocks of saltpetre. Imports from China became a possibility as Ming trading became more open in the 1540s, when these guns were beginning to be used, but still the Japanese struggled to make potassium nitrate.

The 1559 document from Ashikaga Yoshiteru to Uesugi Kenshin describes the production of gunpowder in great detail, down to the best wood to be used for the charcoal and the proper temperature to heat the mixture. This recipe reveals that the gunpowder was not merely mixed, or the ingredients stored separately, but rather they were combined in slurry in this precise ratio. The slurry was then dried and solidified at as a powder. This granulated gunpowder burned consistently and proved more effective. Gunpowder so mixed was known as 'corned gunpowder,' and was originally developed in Nuremberg around 1450. Corned powder did not fully replace the earlier means of mixing dry powder, which was referred to as 'serpentine gunpowder', until guns had become more powerful in the course of the sixteenth century.

BULLETS

Rifled barrels required bullets made of lead, for the metal was soft and malleable enough to not damage the barrel. Bullets could be manufactured easily, and soft metals such as lead were preferred, although archeological evidence reveals that rock and hardened earth were used as well. Lead proved more difficult to procure than earth or stone, however, and so the samurai of the sixteenth century strove to gain more of it. Kobayakawa Takakage would praise Nomi Motonobu in 1557 for 'bringing lead … which can be used for guns.' *Daimyō* armies did not rely on a standard size, although most bullets were 15.8 to 18.7mm (.67 to .74in) in diameter.

A rifled musket could fire to a range of 1000m (1093 yards), which far exceeded that of bows, or for that matter, the earlier *hiya*, but the limit of a meaningful range was 100m (109 yards), and accuracy was limited to around 50m (55 yards). Even as late as 1775, muskets could not be accurately fired at a greater distance. One British office fighting at Concord in 1775 said that 'one might as well shoot at the moon' than fire at a target 182m (200 yards) away. Japanese training manuals emphasize accuracy over speed of firing (for the logic behind this, see military organization below), but the effective limit of the ability to inflict serious wounds was 50m (55 yards).

That such guns were tested suggests that their advantages were not immediately evident. They were heavy, expensive and took several minutes to load. In both Europe and Japan debates arose as to which weapon, the gun or the bow, was superior. Armies commonly mobilized gunners and archers in equal numbers, as can be documented in armies of Henry VIII in 1544, or for that matter, a contingent of Satsuma warriors fighting in Korea in 1591. The superiority of guns versus bows continued to be debated in England during the 1580s, although in the English case the limited number of archers constituted the crux of the debate, while range and accuracy seems to have been the main basis for comparison in Japan, epitomized by a

A depiction of Watonai Sankan, a half-Chinese half-Japanese pirate shooting a tiger. He occupied Taiwan and purportedly strove to re-establish the Ming Dynasty after their 1644 destruction and became a popular figure in kabuki *plays and woodblock prints. This Udagawa Kunisada print dates from 1840.*

BULLET-PROOF ARMOUR

This suit of *tōsei gusoku* armour, made by Unkai Mitsunao, has a reinforced metal cuirass, which provided greater protection against bullets. Substantial chain also protected the arms, and the flexible *okisode* also provided more protection for the upper arms. The *hō-ate* is extended to protect the face below the eyes, and is known as a *me no shita hō*. The base of the helmet is of an older style, but flexible *shikoro* protect the back of the neck. *Yurugi no ito* allowed the lower part of the armour to be adjusted to the height of its wearer, while the divided skirts (*gesan*) are made of lacquered metal. Lacquering helped inhibit rusting. The chain mail and iron plates were made in the sixteenth century, the helmet was crafted in the seventeenth century, and the chords and braiding date from the nineteenth century.

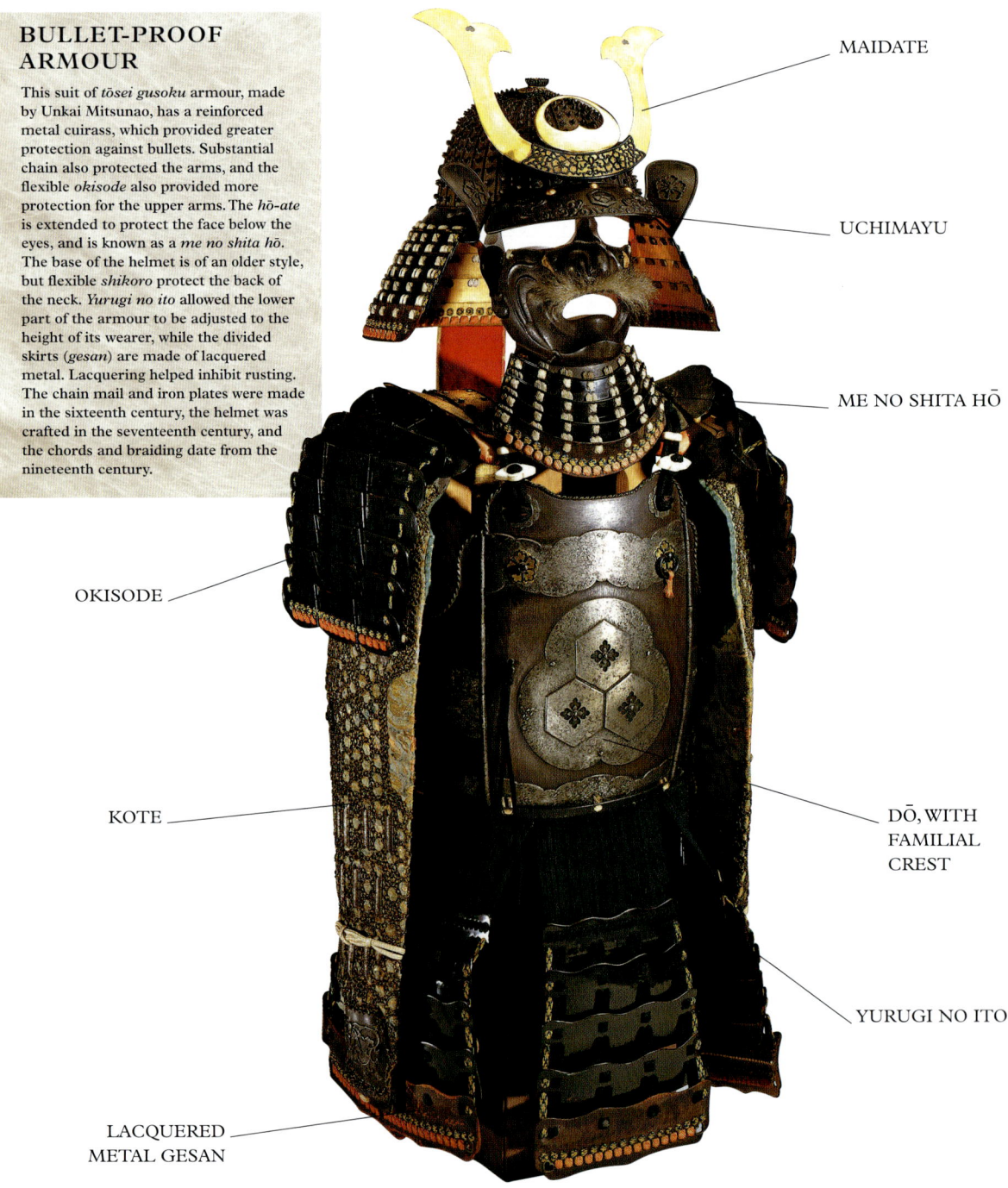

MAIDATE

UCHIMAYU

ME NO SHITA HŌ

OKISODE

KOTE

DŌ, WITH FAMILIAL CREST

YURUGI NO ITO

LACQUERED METAL GESAN

scene from *Famous scenes of the capital (Kyōmeishozu byōbu)* a seventeenth-century screen illustrating life in the capital – an archer and a marksman are portrayed as comparing their weapons during target practice.

Bullets could pierce armour at 30m (33 yards), and therefore kill or seriously wound an opponent. This range represents an almost doubling of the lethal range of projectiles, for as we have seen the Japanese bow could only pierce armour at a range of 12–14m (13–15 yards). As time passed, difficulties in manufacturing these weapons and securing adequate gunpowder receded, and so wounds caused by bullets steadily increased.

FIREARMS

The earliest documented example of a man being explicitly wounded by 'firearms' occurs in 27 January 1563, when Hara Rokurō, a retainer of Sugi Matsuchiyo, was shot near his left armpit by a 'hand-fire arrow' (*tebiya*) by supporters of the Ōtomo in northern Kyushu. Weapons known as 'guns' (*teppō*) (Tanegashima matchlocks based on the Portuguese model) were used to great effect on 13 November 1563, when the Amako of Izumo Province mauled the Kikkawa, wounding 33 by gun, six by arrows, five by rocks and one by a sword.

As a result of Amako prowess, the number of gun casualties supersedes those of bows by a figure of 88 to 64 during the 1560s. Nevertheless, the data is not comprehensive enough to hypothesize about patterns of gun dissemination. Although twice as many bullet wounds (17 to 8) were recorded as arrow wounds in the 1570s, both were inflicted at roughly analogous rates (19 to 16) in the 1580s. Indeed, from 1467 until 1600, arrows caused 58 per cent of all projectile wounds while bullets were responsible for 28 per cent and rocks the remaining 13 per cent. Not until 1600 do surviving documents reveal a pronounced preference for guns, which caused 80 per cent of all projectile wounds on the plains of central Japan.

Guns required nearly half a century to supplant bows because they were only incrementally more effective in range and penetration. Gun wounds were often inflicted in close proximity to enemy forces, as were arrow wounds. Several examples exist of warriors wounded with both bullets and arrows, or being shot and stabbed with a pike in the same encounter. Instances where the same warrior was shot repeatedly and yet survived also attest to the limitations of sixteenth-century firepower.

GUNS AND MILITARY ORGANIZATION

The priests of Negoroji who managed to procure one of three Portuguese harquebuses at Tanegashima, not only manufactured these weapons, but they also quickly and effectively used them. Although it is difficult to know when precisely they organized their gunners in formations, they can be documented as commanding a force of 300 gunners in the 1570s, and these men defeated the brother of Oda Nobunaga.

The Uesugi, for all their lack of administrative proficiency, proved

Samurai discharging firearms while on horseback as depicted in the Budō Geijutsu hiden zue *of 1855. These pistols were designed to be capable of a single, or at the most two shots while on horseback. The inability to reload, and problems in accurately shooting while on horseback meant that this nineteenth-century weapon was not particularly effective unless discharged at exceedingly close range.*

ASHIGARU HARQUEBUSIER (1600)

A marksman. The *ashigaru* depicted here wears simplified armour, and is well equipped to fire his weapon. The considerable amount of equipment required to use guns meant that some *ashigaru* continued to prefer bows until the last decade of the sixteenth century.

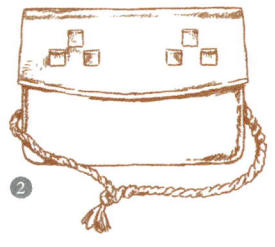

KEY

1. Soft leather bullet pouch
2. Leather cartridge bag
3. Powder flask
4. Storage box for bullets
5. Fuse coil
6. Laquered powder flask with measured stopper

FIREARMS

sophisticated organizers of armies. A document dating from 16 February 1575 reveals that they split into units functionally, according to units of pikes, guns and horsemen. When precisely the Uesugi began organizing their armies into guns is hard to know, but it is noteworthy that shortly after Kenshin received a recipe for gunpowder in 1559 he embarked on a rapid offensive, attacking the Hōjō at their base in Odawara before being invested as 'leader of the east' in Kamakura during the years 1560–61. He also had use of guns at the battle of Kawanakajima in 1562, and inflicted heavy casualties on the Takeda.

Not all warlords used guns effectively. The Takeda were slow in realizing the advantages of firearms. A document dating from 1562 reveals that the Takeda organized their troops differently, with 45 soldiers being mobilized as follows: 30 pikemen, two with short pikes, five archers and one gunner, with the remaining eight serving as porters, or charged with carrying armour, swords or flags.

Only six out of 45 (13 per cent) were capable of shooting projectiles, with 83 per cent of those limited to troops shooting arrows. We can perhaps understand why the Takeda command took such grievous losses at Kawanakajima in 1562. Other *daimyō*, such as the Hōjō, are documented as organizing their armies as the Uesugi in formations of pikemen, archers and gunners, albeit much later, in 1587. They nevertheless relied on archers as an organizational unit more than the Uesugi apparently did.

The Uesugi records also reveal that they were capable of firing their weapons effectively. A document dating from 1615 known as 'the regulations for the Uesugi encampment at Osaka' states that 'regardless of the numbers, guns are to fire in threes'. The Uesugi thus organized their gunners to fire in groups, so as to mitigate the delay in firing. The example of the Uesugi suggests again that those most cognizant of tactical organization, possessed a very real advantage on the battlefield, and this explains why certain families survived and others were destroyed.

In order to best understand this dynamic, let us explore an epic battle, perhaps the most famous battle in Japanese history, which is commonly known as the battle of Nagashino, and witnessed a decisive victory by forces of the Oda and their Tokugawa allies, and the near annihilation of the Takeda.

THE BATTLE OF NAGASHINO

The battle of Nagashino has achieved great fame in world history, for it represents one of the most sophisticated uses of guns in the world; a feat all the more remarkable because of how recently muskets had been introduced to Japan. Geoffrey Parker succinctly

Budō Geijutsu hiden zue *of 1855, showing ingenious methods of holding guns level so as to improve accuracy.*

explains the significance of this battle in his *The Military Revolution* as follows:

> *... everyone admits that the most spectacular demonstration of the power of Japanese musketry occurred on 21 May 1575 at the battle of Nagashino. Nobunaga deployed 3000 musketeers in three ranks in this action, having trained them to fire in volleys so as to maintain a constant barrage. The opposing Takeda cavalry was annihilated. The battle-scene in Kurosawa's film* **Kagemusha** *(The Shadow Warrior) offers a credible reconstruction, for the action is intended to represent Nagashino.*

Parker goes on to explain the significance of this battle, arguing:

> *the originality of Japan's rapid adoption of the gun has perhaps not always been fully appreciated. In the first place, whereas Europe concentrated on increasing the speed of reloading, the Japanese were more interested in improving accuracy … The Tanegashima were, for their day, remarkably accurate. But this in fact accentuated the crucial defect of the muzzle-loading musket: the length of time required to recharge it. As noted above, the only way to overcome this disadvantage was to draw up the musketeers in ranks, firing in sequence, so that the front file could reload while the others behind fired. This solution was not even suggested in Europe until 1594, and it did not pass into general use there until the 1630s. Yet Oda Nobunaga had experimented with musketry salvoes in the 1560s, and he achieved his first major victory with this technique in 1575, twenty years before the European innovation.*

Oda Nobunaga, of a deputy *shugo* family, was famous for 'unifying' Japan, and bringing the beginnings of a new and brutal political order. His innovation was legendary, as too was his military prowess, and the degree of his success has thought to largely stem from his use of weapons.

The battle of Nagashino represented a rare case of an utter rout of the Takeda by the Oda, which hastened their ultimate destruction in 1582. Takeda Shingen had been succeeded by one of his sons, Katsuyori (1546–82), who has gone down history as one of the least successful of 'Warring States' commanders. Akira Kurosawa's film *The Shadow Warrior* depicts a suicidal charge by the Takeda, who under the poor command of Katsuyori, senselessly

FIREARMS

charge their pike and horsemen into heavily fortified barriers manned by gunners who almost completely annihilated their forces.

For all of its fame, the battle of Nagashino has not been particularly well analyzed. Owada Tetsuo, in his *Nagashino Shitaragahara no tatakai*, has complained that much of the analysis of this battle has come from individuals who knew nothing about the battlefield, or had never travelled there. But the battlefield remains intact, and the author had the opportunity to visit this site in June 2005. An exploration of battlefield terrain, when combined with analysis of archeological evidence and surviving

LEFT: *A scene of horsemen galloping from Akira Kurosawa's* The Shadow Warrior *(*Kagemusha*). This 1980 film's depiction of Nagashino is based upon seventeenth-century screens of the battle.*

BELOW: *The centre of the Nagashino battlefield, facing north. Tokugawa forces were entrenched on the left, while the Takeda would have been in the vicinity of the house on the right. The creek separating these forces is faintly visible as a depression in the center of the field. The right wing of the Takeda army occupied the hills visible at the centre of this image.*

documents suggests that harquebuses contributed to the Takeda defeat, but that their tactical miscalculations proved decisive. Katsuyori lost because he unsuccessfully attempted to encircle Tokugawa Ieyasu's army, and had one wing of his army cut off instead.

THE TERRAIN AT NAGASHINO

The battle of Nagashino has been misnamed, for Nagashino was a nearby castle and did not constitute the focal point of the battle. This castle, besieged by the Takeda, had not fallen before reinforcements arrived, drawn from the armies of Oda Nobunaga and Tokugawa Ieyasu. As the Oda and Tokugawa armies hurried east in order to break the Takeda siege, Takeda Katsuyori turned and marched his army 2km (1 mile) to the west to confront them at a place called Shitaragahara. At was here that the Takeda launched an attack, and died.

Shitaragahara is less rugged than one might assume upon reading topographical maps of the region. The focal point of the battle is the low rise of the Danjō Hills, overlooking a small creek called the Rengo River. The hills rise and end abruptly, leaving open plains to the south that were, however, guarded by a castle. These hills gradually

LEFT: *Tokugawa Ieyasu, who fought at Nagashino, and later founded the Tokugawa regime after triumphing at the battle of Sekigahara. He was later deified as the* Tōshō Gongen, *or 'Shining Avatar of the East'.*

recede to the east as one travels to the north, and one perceives a break in the hills, which in fact does not exist. Immediately across from what appears to be a break in the hills, and across the Rengo, there is a hill called Maruyama.

The Takeda and Tokugawa forces had been stationed along the Rengo River, but Nobunaga, occupying the southernmost line on the plains, withdrew his army to the rear of Tokugawa Ieyasu's, making his encampment at Cha'usu Mountain. It is noteworthy that he evacuated an area of clear plains for a rear position – seemingly ideal territory for the Takeda to envelop the Tokugawa line. The Tokugawa forces occupied Maruyama Hill, which

FIREARMS

ABOVE: *View from the reconstructed barricades located at the centre of the Tokugawa lines looking towards the Takeda, who would have been stationed in the hills just beyond the Rengo Creek, which appears as a depression in the centre of these paddies.*

CENTRE LEFT: *A view of the northern part of Nagashino, from the Takeda lines. The Takeda took control of Maruyama, the hill located to the right of this photo, and then part of their armies advanced in the apparent gap behind this hill, where they were cut off and annihilated. From this vantage point, they did not realize that substantial Oda reinforcements were hidden behind the Danjō Hills.*

LOWER: *View of the Danjō hills from the Takeda line. At this distance, some Takeda generals suffered casualties from Tokugawa and Oda guns. The extensive barricades, reconstructed here, most likely did not exist, for the Tokugawa were protected by the Rengo River to the fore and the Danjō Hills to the rear. Instead, barricades most likely were placed to the south, or in the gap beyond Maruyama.*

apparently controlled access to a gap in the Danjō Hills.

The current placement of bamboo barricades, added to the battlefield in the twentieth century, simply does not make any sense, for they are located at the centre of the Danjō Hills. The Takeda would not have crossed the Rengo and attempt a frontal assault on these small, but steep hills. Analysis of written and visual sources, when coupled with an exploration of archeological evidence suggests that the Takeda never attempted a major assault on the Danjō Hills behind these barricades at all.

THE EVIDENCE

Many elements of the battle remain little understood. Sources pertaining to this conflict are limited, and contradictory. Debates have raged over the phrasing of certain passages, or for that matter the veracity of written sources, but terrain and archeological evidence have not been adequately explored.

No one knows how many muskets Nobunaga and his Tokugawa allies employed in battle. The number 3000 is commonly bandied about, but this figure is not reliable. The *Shinchō kōki*, a remarkably reliable chronicle written in the late sixteenth century by a close Oda

TEPPŌ GASHIRA

GUNNERY CAPTAIN

Commander, or literally 'head' of a gunnery unit (*teppō gashira*). Units of marksmen did not exist until late in the sixteenth century, and the earliest reference to such a commander comes from 1585. The head of these units possessed a hollow baton, which could be used as a ramrod if necessary.

AMMUNITION CARRIER

This depicts a man carrying a box of ammunition to battle. Bullets and gunpowder were in short supply and had to be kept dry. This man wears a sword but others were presumably unarmed, like most porters.

AMMUNITION BOX

shipment proved as significant, if not more so, than the appearance of the 500 or so gunners who joined the Oda forces.

None of the oldest, and thus most reliable accounts, state than gunners fired in groups of three. Uesugi records show that this was common practice by 1615, still very early in the span of firearms history, but it seems unlikely or at least impossible to verify that gunners fired in groups of three in 1575.

References to the famed Takeda cavalry charging are also absent from contemporary sources. As we have seen in chapters 3 and 4, sixteenth-century cavalry was more used for reconnoitring than charging, and horsemen generally consisted of 10 to 20 per cent of an army. It seems unlikely that the Takeda were fundamentally different from their rivals, and so their forces must be supporter, originally described the Oda forces as having 1000 gunners, but someone later revised this figure to 3000. A smaller force is likely. Other evidence suggests even fewer gunners than mentioned in the *Shinchō kōki*, for Nobunaga in a letter to Sakai Tadatsugu, a retainer of Tokugawa Ieyasu, on the night before battle, describes a force of 4000 which included, among them 500 gunners. This figure is close to that used to describe the force of 300 gunners of Negoroji, who proved effective at this time.

Less than a week before the battle, Nobunaga, in a letter to Hosokawa Fujitaka dated 15 May 1575, acknowledges receiving guns and gunpowder from the vicinity of the capital. The difficulty of manufacturing high-quality gunpowder meant that this

FIREARMS

ASHIGARU ARCHERS

Even in the armies of the sixteenth century, archers continued to be used on the battlefield as their weapons were cheaper than guns and they could fire more rapidly. Often they fought side by side with marksmen. The crest of the archer's *daimyō* is depicted prominently on their armour.

JINGASA

WAKIZASHI

primarily visualized as fighting on foot.

The next point of clarification concerns the placement of the barricades. The current reconstruction of the barriers is nonsensical for there was no need to build barricades behind rivers and in front of hills. Close analysis of extant screens, reveals, however, a more sensible distribution of the barricades. The Nagashino battle screens owned by the Nagoya Museum are the oldest representation of battle, dating from the 1590s, and thus well within living memory of the events that had occurred. These screens depict the disposition of the Tokugawa and Oda forces and shows that they are stationed behind relatively fragile barriers.

Significantly, these barricades guard fields to south of the Danjō Hills, where a flanking manoeuvre could be attempted, and to the north of Danjō as well. To the north, a few gunners are depicted in front of a barricade, and then the Oda forces are hidden behind an apparent gap in the land, and mountains. Another, more famous and relatively accurate screen depicts the dispositions in a similar manner. This work, owned by the Narase family, depicts four larger, more durable walls. The first was to the plains to the south of Danjō, a logical place for Tokugawa Ieyasu's forces to be outflanked. A second barricade existed in front of Danjō, to presumably protect Ieyasu, and then another barricade, opposite of Maruyama, where an apparent opening existed in the line. Behind this region, one sees a second set of barricades, the existence of which has not been commented upon much, as from the scrolls, it appears that none of the Takeda were able to break the initial line of barricades that were set up by the Oda and Tokugawa.

Archeological evidence both supports and undermines these dispositions. A total of 11 bullets have been discovered on the battlefield. Six are located in front of Danjō Mountain, where the two armies fought. Significantly, two bullets were likewise uncovered to the south, emblematic of a flanking attack. The remaining three were discovered, remarkably, on the western side of Danjō Mountain. This location is well behind the Tokugawa lines, and somewhere near where the Oda forces were hiding in reserve.

THE BATTLE OF NAGASHINO: A RECONSTRUCTION

The battle began at 6am, and in the following hour, to drum beats, the forces of Yamagata Masakage and Naitō Masatoyo, located on the far left of the Takeda line, led a charge against forces

WEAPONS AND FIGHTING TECHNIQUES OF THE SAMURAI WARRIOR

barricades kept these forces from breaking through. The existence of a castle, however, to the very south, and the very obviousness of this as a route of attack suggests that this area was well defended, and it is unlikely that the Takeda saw this as their key to victory.

Takeda commanders suffered disproportionate casualties. At least 37 prominent warriors perished. The Takeda were aware of guns, having had some for nearly 50 years, but it seems unlikely that they had been exposed to the best guns, or the most potent gunpowder. Takeda commanders were surprised by the range of the new guns, which as we have seen, had killing potential of approximately 30m (33 yards), far longer than arrows

Nagashino battle screen. This version, created during the seventeenth century, is the most famous, but less accurate than the earliest depictions, dating from the 1590s. Akira Kurosawa's battle scene is based on this screen, and the modern placement of barricades on the current battlefield is drawn from this image as well.

occupying the southern fields below Danjō, attempting to flank the Tokugawa forces to the south. The forces of the Ōkubo Tadasuke and Tadayo stopped this offensive, and both Takeda commanders lost their lives.

Repeated feints here produced few results. The supporting army in the centre held firm and took heavy casualties, presumably to buy time for the feint around the right of the Tokugawa line to succeed, but

DEFENSIVE POSITIONS

Marksmen required so much time to reload their weapons that they were vulnerable to attack, and so hid behind shields, trenches and other barricades. Although only gunners are depicted here, they would have also been stationed with archers and pikemen for added protection.

or the more primitive guns. As commanders wore readily identifiable armour, they would have been great targets, particularly when they were used to directing units at close range, and so were surprised at this improvement in infantry firepower.

Overall Takeda fatalities appear to have been at approximately 17 per cent, a rather high number, for a contemporary courtier diary recounts that 1000 were killed, while Takeda forces have been estimated as consisting of 6000 soldiers. By contrast, the Oda-Tokugawa forces seem to have been much greater, approaching, to the best estimate, 18,000 men, but their casualty rates are not known.

The Takeda attempted to envelop the Tokugawa from the north, and this catastrophic failure led to the death of many of their men. The right wing of the Takeda line succeeded in crossing the northern line of the Danjō Hills. Under the command of Baba Nobuharu, Tsuchiya Masatsugu, Anayama Nobukimi, and Ichijō Nobutatsu they took Maruyama, and then advanced most successfully here, fighting against the Oda commander Sakuma Nobumori.

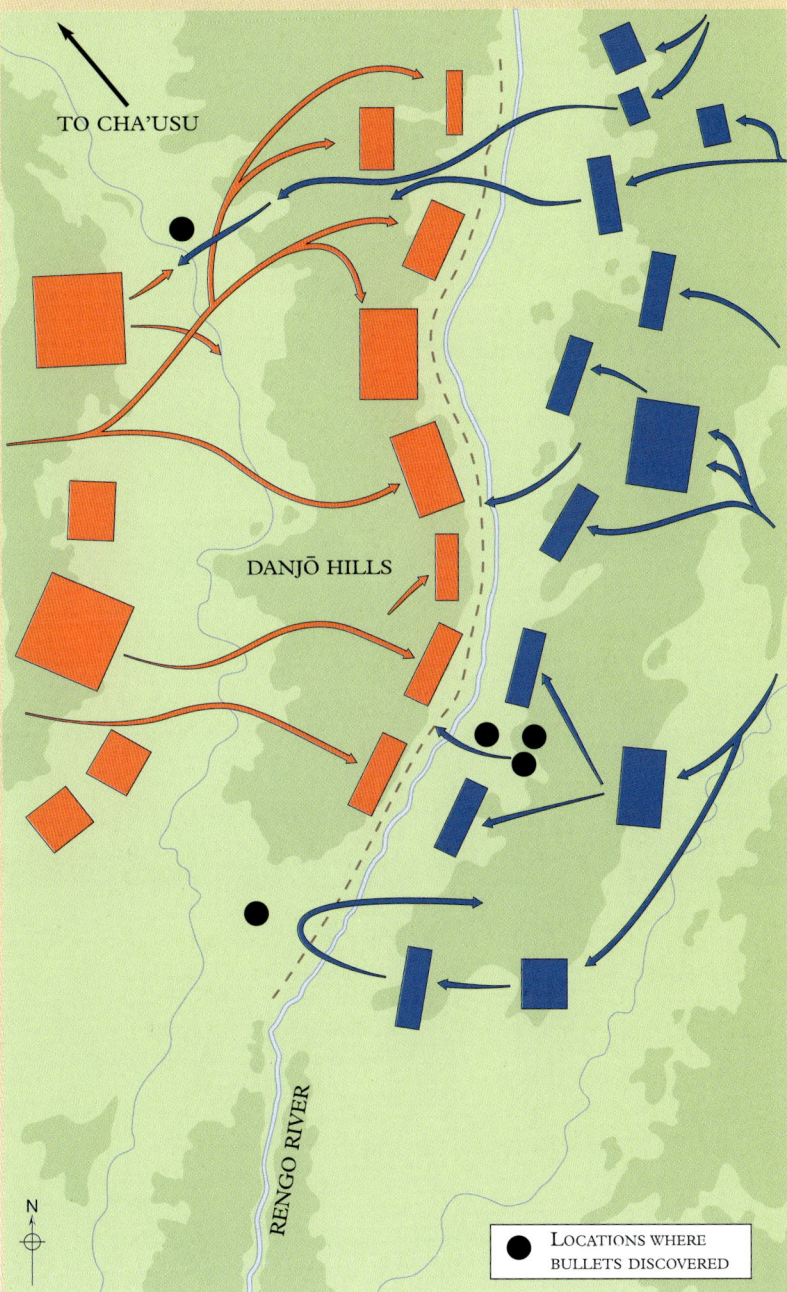

NAGASHINO, 1575: THE CHARGE
The Takeda (blue) attacked the combined Oda and Tokugawa forces (red), and attempted to envelop them. Rumours of treachery encouraged the right wing of the Takeda to cross the Tokugawa lines, where they were then surprised by concealed forces and annihilated. The left wing was turned back as well, while the centre of the Takeda line suffered casualties in their attempt to hold the Tokugawa in place.

WEAPONS AND FIGHTING TECHNIQUES OF THE SAMURAI WARRIOR

A nineteenth-century portrait of the famous warlord Oda Nobunaga (1534–82). Oda Nobunaga lived a life of continuous military conquest, conquering a third of all Japanese daimyō before his death in 1582.

Nobumori, in a letter that he later wrote, explained that he had sent a secret messenger to the Takeda, informing them of his intent to defect from the Oda side, and so the attack on this section of the line proved powerful. The Takeda commanders expected Sakuma to turn, thereby allowing them to roll up and surround the forces of Tokugawa Ieyasu from the north.

Unaware that they were being led into a trap, the Takeda commanders apparently broke through the Oda line of barricades here and, from the perspective of the other forces on the field, passed out of sight. The Takeda expected that this force would cut off the Tokugawa, which is why the left and centre of the line continued to engage and hold the Tokugawa, so as to allow the envelopment to be complete. But the advancing forces encountered Oda Nobunaga's large, hidden and entrenched forces, and were annihilated. Two of the four commanders who led forces on the right side of the line – Baba Nobuhara and Tsuchiya Masatsugu – were killed. Other warrior families who can be documented as fighting on this side of the line, such as the Saneda and Tsuchiya, suffered multiple deaths as well. These men were killed on the right side of the line bcause they had no support, and were shot by Sakuma and endured the brunt of the Oda counterattack. To make matters worse for the Takeda, by the time their failure became known, the rest of the line had endured much punishment, and began a general retreat. This is the most dangerous manoeuvre, for as soldiers flee they are even more vulnerable to being killed, which then caused casualties to mount further.

SOCIAL LEVELLER

Guns contributed to the high casualties among Takeda officers, but it was Oda Nobunaga's cunning ability to hide his army, and deceive the Tokugawa, that caused such an epic defeat. By 10am, the battle had been decided, and the surviving Takeda began their retreat.

Both commanders to the south, Yamagata and Baba, had been killed and in addition commanders to the right of the line suffered grievously. The Tsuchiya lost two of their leaders, and Anayama Nobukimi perished as well. Takeda Katsuyori, all too belatedly realized that he had been led to attack by a ruse, and hopes of surrounding an annihilating Tokugawa forces ended with the crippling of his own instead. The defeat proved decisive, and led Nobunaga to write letters to various *daimyō* in celebration, among them a very happy Uesugi Kenshin.

Guns were significant in that they caused high- and mid-ranking Takeda commanders to suffer heavy casualties. That so many would fall stems in part, perhaps, because the Takeda were less familiar with the power of the new matchlocks, although Uesugi Kenshin can be documented as having new guns and highly explosive powder during the fourth battle of Kawanakajima in 1562, another place where the Takeda again suffered disproportionate casualties among their commanders. Shingen avoided a devastating defeat in 1562 because he, unlike his rash son, did not allow one wing of his army to be enveloped and annihilated.

Takeda Katsuyori, who again praised his followers for providing gunpowder before the battle, proved more insistent on the need for such powder and guns immediately thereafter. Writing in the November 1575, Katsuyori would, in new regulations, state that 'now guns are most essential'. Whereas before, he had praised those who brought guns instead of pikes, now he was more specific, asking that those who brought guns also provided skilled marksmen from among the *ashigaru*. Also, alluding to a startling weakness of the Takeda, particularly when compared to the Uesugi and, presumably, Oda and Tokugawa forces, he stated that those who could not fire a gun or bow should not show up in camp, revealing that his armies had too many camp followers and not enough skirmishers. So pressing was Katsuyori's need for more gunners and archers that he stipulated that all soldiers reporting to camp would be inspected upon their arrival, and those without adequate bows or guns would be declared negligent.

After the debacle of Nagashino, Katsuyori paid more attention to the composition of his army. All soldiers were to wear armour (*gusoku*) a helmet, gloves, a throat guard (*nodowa*) and banners. Gunners, too, would be required to bring 300 rounds of gunpowder and ammunition per rifle, which suggests that such supplies had previously been insufficient.

According to nine documents, one can verify a total of 94 men being

Gunners wearing sixteenth-century armour, with the man in the foreground showing the Oda family crest, firing muskets as part of the teppō hyakuninchō *contingent of Tokyo's Kaichū Inari shrine, 25 September 2005.*

mobilized during 1576. This force was composed as follows: eight cavalry, 10 flags, 13 guns, 11 archers, 21 medium' pikes of a length of 6m (18ft) and 31 long pikes. Cavalry only constituted 8.5 per cent of the army, with the remaining 91.5 per cent were on foot. Archers and gunners were mobilized at nearly equal rates: 13 to 11, and thus those who relied on projectiles constituted a quarter

MAEDA TOSHINAGA

Maeda Toshinaga (1562–1614) took over the Maeda domains after his illustrious father Toshiie died in 1600, and supported Tokugawa Ieyasu at the battle of Sekigahara the same year. He is depicted here wearing an *eboshi* shaped like a giant catfish tail (*namazu o nari kabuto*). The helmet, being made of black laquer, is actually surprisingly light. Toshinaga's crest is visible on his gauntlets.

ABOVE: Protective boxes to keep the firing mechanisms of Japanese harquebuses dry, from the Budō *Geijutsu hiden zue of 1855.*

(25.5 per cent) of the army. The majority of the army (55 per cent) continued to consist of pikemen, although this figure represents a decrease from 1562, when 71 per cent of all Takeda troops wielded pikes.

Katsuyori required more skirmishers, and seems to have favoured archers and marksmen equally, suggesting that the need to fire projectiles mattered more than the weapon used per se. In 1562, however, only 13 per cent of the army (six out of 45) specialized in shooting, so by 1576 that figure had now doubled. The ratio of guns to arrows shifted from 17:83 in 1562 to 54:46 in 1576. The number of guns tripled in 14 years, and they became slightly more common than bows. That guns and bows were used equally can also be verified in casualty lists from the 1570s and 1580s, as both inflicted wounds at comparable rates. Likewise, a roster in the *Documents of Iriki* collection dating from the Korean campaigns of 1591 reveals that archers and gunners were used in equal numbers – 1500 each – in Shimazu Yoshihisa's army of 15,000.

MASSED MUSKETRY

Muskets were able to achieve parity with arrows, but they could only become an effective weapon when gunners were organized into cohesive groups, a process which only began during the 1570s. With the increased numbers of gunners

During the sixteenth century, commanders tended to take a more active role in fighting in campaigns. This scene from the Akira Kurosawa movie Ran *shows a higher ranking samurai on an impossibly large (by contemporary Japanese standards) horse.*

incorporated into armies, they were organized into distinct groups of pikes, gunners and horsemen by the Uesugi on 16 February 1575, several months before the battle of Nagashino. They are documented as using a variety of guns, for 24 bullets of varying calibre have been uncovered at Kasuga Castle, dating from a disturbance in 1578.

The leaders of eastern Japan seem to have laboured under a comparative disadvantage when compared their western cohorts. Those with good ties to the capital, such as the Uesugi, fought well, but the Takeda and indeed the Latter Hōjō appear to have been much less effective.

Hōjō mobilization reports from 1581 reveal that in one case, only 14 per cent of one unit (eight out of 56) could fire projectiles, with bows being more common than guns by a ration of 5:3. Another retainer only had one archer and one gunner out of a force of 26 men. Not until 1587 does this ratio increase to 28 per cent of the army using projectiles, with guns and arrows used at an equal ratio (20:20). It was at this moment that Hōjō Ujimasa referred to formations of pikemen, bowmen and gunners in a 1587 report, and likewise in a document dating from 1588 one can see that the Hōjō organized their gunners in groups of ten men.

Knowledge of guns seems to have improved considerably over the course of the 1580s, as a 1585 treatise on bullet types and their efficacy attests. Furthermore, documents written during Oda Nobukatsu's Komaki Nagakute campaign of 1584 reveal that a force of 50 gunners proved capable of inflicting substantial casualties.

The Hōjō were destroyed by the forces of Japan's second unifier Hideyoshi, who succeeded Oda Nobunaga after he had been assassinated by one of his generals, Akechi Mitsuhide, in 1582. Shortly before his death, however, Nobunaga managed to annihilate the Takeda, and his successor Hideyoshi would destroy the Hōjō in 1590. Although good battle records do not exist, it seems that over the course of the 1590s use of guns shifted dramatically. Guns did not fully displace bows until the year 1600, when they inflicted 80 per cent of all skirmishing casualties. Likewise the pike remained the dominant shock weapon in battle, inflicting over 98 per cent of hand-to-hand wounds as well.

Even as late as 1600, not all commanders realized how vulnerable they were to marksmen, in spite of the lesson that Nagashino had purportedly offered. For example, during the battle of Sekigahara, the Shimazu had sided against the Tokugawa and were betrayed and their allies defeated. Ii Naomasa (1561–1602), a loyal *daimyō* of the Tokugawa wearing prominent armour, commanded his troops to destroy the surrounded Shimazu, but at that moment he was shot. Naomasa did not die, but in the confusion the Shimazu managed to cut through Tokugawa lines and escape.

Although Japanese guns have attracted considerable attention, they did not alter the nature of battle as much as allow men to fire projectiles at twice the distance than had been common before. This caused an upswing of casualties among high-ranking warriors wearing

Negoroji Temple. Negoroji priests manufactured and used firearms until Toyotomi Hideyoshi mobilized a large army, which overwhelmed the Negoroji's skilled gunners. The two surviving structures have bullet holes of varying sizes, suggesting the different calibres of guns used in the attack.

flamboyant armour, as the Takeda suffered in 1575, but the nature of battle did not change. Massed units of pikemen continued to occupy the battlefield. Pikes witnessed an increase in length to 8.2m (27ft), which reveals that warriors tended to fight at a greater distance from each other than before.

The existence of guns gave warriors from western Japan, and those like the Uesugi with ties to the capital, an advantage over their more eastern rivals. Save for the Uesugi, most of the major warrior *daimyō* of the east – the Imagawa, Takeda and the Hōjō – met with overwhelming military defeat and were destroyed over the course of a generation. Improvements in military organization, and the widespread use of guns led to a change in the military balance, as massive armies now proved capable of overwhelming even small but skilled groups of gunners. The priests of Negoroji discovered this to their detriment in 1585, when Toyotomi Hideyoshi burned all but three buildings of their temple. The three surviving Negoroji buildings are peppered with bullet holes, again revealing the prevalence of these weapons.

Nevertheless, the story of the establishment of peace, and the creation of a new order of Tokugawa hegemony deserves analysis of another weapon – the cannon – and an exploration of how fortification techniques changed, and finally how an under-examined relationship between the samurai and these weapons would shape the course of the next three centuries.

LEFT: *The battlefield of Sekigahara, where treachery by the Kobayakawa allowed the 'Eastern Army' of Tokugawa Ieyasu, which was outnumbered and in inferior position, to defeat the 'Western Army' of Ishida Mitsunari and secure Tokugawa military hegemony for the next two and a half centuries.*

SHIP-TO-SHIP COMBAT

Although little is known about these boats, heavily armoured craft were apparently used by Oda Nobunaga, in his attack on Ishiyama Honganji, while a famous 'Turtle ship' was used by Korean Admiral Yi Sunsin against Toyotomi Hideyoshi's invasions of 1592. These craft were not speedy or agile, and instead were platforms for pikemen, archers and gunners to fire on their opponents.

CANNON AND ARTILLERY

Cannon have been almost completely overlooked in coverage of the dissemination of guns to Japan. This lacuna proves surprising, because the Portuguese came in ships bristling with cannon, and these weapons were coveted by *daimyō*. In contrast to the firearms, which were impressive but continued to be used in equal numbers with bows for much of the sixteenth century, the superiority of the Portuguese cannon to earlier artillery, such as the catapult, was immediately obvious.

Detail of part of a folding screen depicting the siege of Osaka Castle (1615). Gunpowder weapons were widely used at this siege, which is indicated by the great puffs of smoke in the illustration.

Later samurai attitudes have contributed to the obscurity of cannons. Literary accounts of battle, and summations of samurai ideals, emphasize the role of individual valour and honour. Few treatises existed glorifying cannon as being weapons of the samurai. Cannon were weapons destined for armies, and little glory accrued to their use.

Furthermore, the possession, production and use of cannon were all tightly controlled by the descendants of Tokugawa Ieyasu, who had successfully used them to bludgeon his rivals into submission. Fearing that imported weapons would alter the military balance of power, the Tokugawa resorted to monopolizing the production of these weapons and firearms, and also limited contact to Europe via one port, Nagasaki, where the Dutch resided on the tiny harbour island of Deshima. Their 'closure' of Japan helped the Tokugawa to preserve their monopoly on the importation of artillery pieces until the early part of the nineteenth century.

ASIAN ARTILLERY

Little is known about artillery in Japan prior to the introduction of cannon. The earliest mechanisms were catapults, which were used in the fourteenth century. According to the *Taiheiki*, Kusunoki Masashige used one while besieged by Aso Harutoki's army in 1333, but of this weapon little is known. Masashige's castle of Chihaya occupied a steep and inaccessible mountain, so whatever catapult he may have used must have been relatively small.

The first verifiable account of a trebuchet comes from the time of the Ōnin War in 1468, when, as we have seen, a trebuchet capable of launching a 3kg (6.6lb) projectile for over 274m (300 yards) was mentioned in the chronicles. A specialist manufactured this catapult, for the account mentions that a craftsman from Izumi Province built it, but of its specifications little is known. The size of the projectile mentioned, however, is much smaller than European trebuchets of the fifteenth century, suggesting that it was able to damage prominent watchtowers behind enemy trenches, but was not capable of smashing holes in rock defences.

EARLY CANNON

Primitive cannon existed in the fifteenth and sixteenth centuries in China and Korea, and these were smoothbored weapons. They were not, however, particularly powerful, for reasons of craftsmanship and the fact that, as we

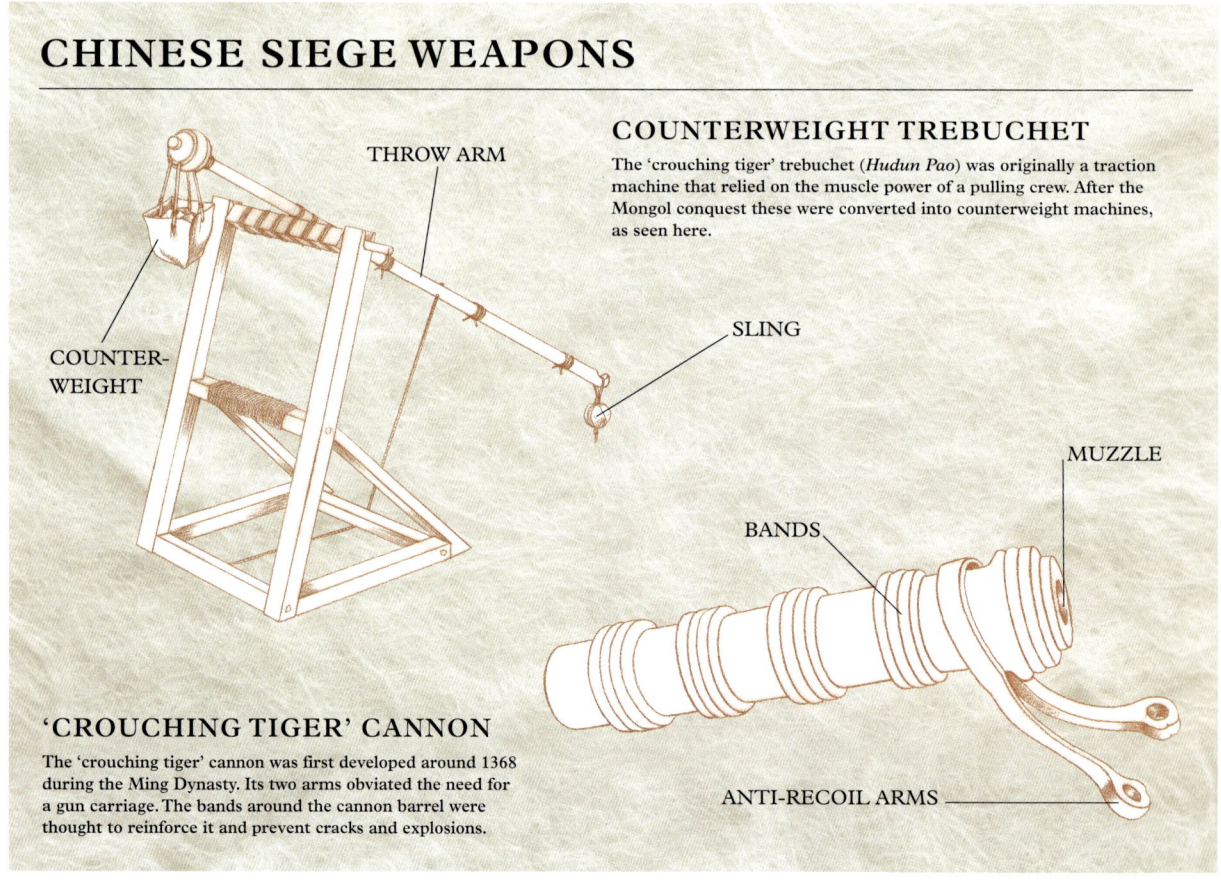

CHINESE SIEGE WEAPONS

COUNTERWEIGHT TREBUCHET

The 'crouching tiger' trebuchet (*Hudun Pao*) was originally a traction machine that relied on the muscle power of a pulling crew. After the Mongol conquest these were converted into counterweight machines, as seen here.

'CROUCHING TIGER' CANNON

The 'crouching tiger' cannon was first developed around 1368 during the Ming Dynasty. Its two arms obviated the need for a gun carriage. The bands around the cannon barrel were thought to reinforce it and prevent cracks and explosions.

JAPANESE CATAPULT

No examples of Japanese catapults survive, even though they can be documented as being used to great effect in 1468, as they fired projectiles against enemy watchtowers. These trebuchets probably resembled the 'crouching tiger' Chinese versions, and as the projectile fired was relatively small, they were not the massive structures such as appeared in contemporary Europe. Unfortunately, the craftsmen who manufactured these weapons did not leave any diagrams of these contraptions.

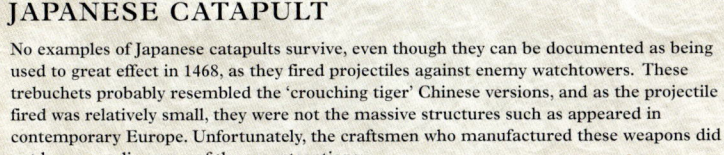

have seen, the Chinese did not have the optimum recipe for gunpowder. The earliest bronze cannon, powered by saltpetre-rich powder, proved to be far more effective.

A variety of names exist for these early cannon. Many, located on ships bows, were swivel cannon. These breech-loading weapons were commonly called *furanki*, which alludes to the 'Frankish'

A ribbed cannon on the parapet of the Great Wall of China. Chinese cast-iron cannon had been used since the mid fourteenth century, but their quality was poor. They could not be fired safely until the latter half of the sixteenth century. High quality guns continued to be manufactured from bronze until the early nineteenth century.

CANNON MANUFACTURE

First, a steel plate was rolled to make a barrel, which was reinforced by wrapping steel strips around it. After that the barrel was melted, shaped and one end was closed off. The thickness of the steel contributed to the strength of the barrel.

STEEL PLATE

WRAPPING THE STRIPS AROUND THE ROUNDED STEEL PLATE

ROLLING THE PLATE

COMPLETED WRAPPED BARREL

HAMMERING IT INTO A TUBE

RESHAPING THE BARREL, AND CLOSING OFF THE END

A SECOND STEEL PLATE FOR CUTTING INTO NARROW STRIPS

COMPLETE GUN BARREL

Some examples of these early breech-loading weapons survive. These *furanki* are made of bronze, and weigh approximately 120kg (nearly 270lb). The breech cover itself was made of wood, and had the disadvantage that some origins of these guns, even though it was the Portuguese who introduced these weapons to East Asia. Breech-loading weapons were also referred to as *ishibiya*, or 'stone fire arrows', a term that contrasts with the complementary term for firearms (*tebiya*), which quite literally means 'hand fire arrows'. Muzzle-loading weapons were referred to as *hakaran*, although the etymology of this word remains obscure. Finally, cannon also became known as 'country destroyers' (*kuni kuzushi*).

Furanki were manufactured as early as 1523 in China, but the Portuguese seem to have monopolized the construction of the very best of these weapons. Ōtomo Yoshishige (Sōrin) first gave a cannon (*ishibiya*) to Ashikaga Yoshiteru in the third month of 1560, and shortly thereafter he requested another from the Portuguese. He wrote: 'Please send another cannon, as I live on the coast, near the border with enemies, and I need to protect myself from them. If I can protect my domains with a cannon, and these lands can then prosper, I will build a church for Deus and allow Padres and Christians to travel forth, and finally, I will allow the Portuguese to reside here.' This account reveals the close connection between missionary work and the dissemination of cannon.

RIGHT: *Samurai and* ashigaru *firing a cannon from the 1855* Budō geijutsu hiden zue.

explosive gas escaped along its seams, causing the distance and velocity of its projectile to decrease. Still, these guns could fire a 70mm (2.76in) projectile that weighed 1.3kg (2.87lb), and the breech arrangment meant that these

SIXTEENTH-CENTURY CANNON

PORTUGUESE BREECH-LOADING CANNON

These cannon, known in Asia as *furanki* ('Frankish') were breech-loading weapons, with a swivel for mounting on ships. The charge and cannonball was placed in the open breech, which was covered with a piece of wood, allowing the cannon to fire rapidly. The chamber closed with a breech cover (this part has now been lost), which was held in place with a peg of wood or iron. The breech cover allowed some explosive gas to escape, thereby lessening the power of the explosion, and the velocity of the projectile.

BREECH CHAMBER

IRON TILLER

SWIVEL MOUNT

AKUNE CANNON

This was discovered off the coast of Kyūshū in 1957, and is thought to be one of the earliest Portuguese cannon that was brought to Japan.

BREECH CHAMBER

weapons could be reloaded quickly, and need not be cleaned out laboriously between shots like later, more powerful versions. Many were also mounted on swivels for ease of movement, and were a rapid firing, effective weapon on ships.

Only the most powerful of western *daimyō* managed to receive a small number of these guns. In addition to the Ōtomo of northern Kyūshū, the Mōri of western Japan, who succeeded the Ōuchi, owned such a weapons, for a detailed drawing remains in their archives, although the original no longer exists. The Shimazu of Satsuma possessed them as well. One specimen was discovered 100m (109 yards) off the shore of Akune City, in Kagoshima (formerly Satsuma) Prefecture in 1957.

These cannon were, because of their weight, best suited for defending castles. According to lore, the Ōtomo had two cannon secretly brought to their Ni'u castle in Usuki, and when the Shimazu attacked, the defenders bombarded them, thereby forcing the Shimazu to abandon a four-day siege of the castle. These weapons seem to have inflicted great casualties on the Shimazu attacking Ni'u.

Macao cannon. The great bronze Portuguese cannon forged in Macao in the early seventeenth century were so well made that they continued to be used into the early nineteenth century.

> ... as a number of cannonballs entered Ryūzōji dwellings, the screams of the women and children could be heard as far as the distant Arima encampment. Yamada O'an, the daughter of Yamada Kyoreki, endured a bombardment in Ōgaki castle in Mino, would comment in her diary that "when cannons fire, the keep [yagura] of the castle sways, and the earth sounds as if it is being rent asunder".

Still, cannon suffered from drawbacks – as the Ōtomo discovered when they advanced on the Shimazu in Hyūga with their two cannon. They were very heavy, difficult to transport and used a considerable amount of precious gunpowder. Something of the difficulty of transport appears in a letter written by Ōtomo Sōrin sometime between 1560 and 1570, immediately after he had received a second cannon from the Portuguese, for he wrote: 'Cannon [*ishibiya*] arrived at the harbour of Takase. It would be good if you could quickly dispatch porters – a great number of them are needed – so send them my way quickly.' At the 1578 battle of Mimigawa, Sōrin, flush with victory, had advanced to wipe out the Shimazu, but suffered a devastating defeat. Unable to use his cannon effectively, he abandoned what Luis Frois described as 'exceedingly excellent cannon' to his Shimazu enemies.

Cannon came to be more commonly used in battle. Arima Harunobu (1567–1612), an ally of Shimazu Iehisa (1547–87), fought Ryūzōji Takanobu in 1584, and trained their cannon on his location in an attempt to kill the enemy commander. The Jesuits would comment on how these two smallish weapons terrorized the Ryūzōji, for:

It is difficult to know when cannon were first manufactured in Japan. Some believe that this began as early as the 1570s. The *Kunitomo Teppō ki* reveals that as early as 1571 Nobunaga ordered craftsmen from Kunitomo Village to construct two guns of 3m (9ft 10in) in length and firing projectiles of 74g (2.61oz). Kunitomo began

CANNON AND ARTILLERY

manufacturing guns from the mid sixteenth century, and when the area of Nagahama where they resided entered the control of Nobunaga, they continued to forge weapons for him.

When making cannon, the Kunitomo smiths first rolled a central cylinder and then several layers of steel plating would be wrapped around it, and then the completed gun would be melted again, and finally shaped. The guns manufactured proved sturdy and they could become extremely thick, which did little to help their mobility but ensured that they would be durable.

Large weapons continued to be manufactured in Sakai by the Inadome family. These muzzle-loading weapons, known as 'cannon' (*taihō*), consisted of a thick body, often protected by wood. In contrast to the *furanki*, which fired a smaller projectile of slightly over 1kg (2.2lb), some of these weapons were capable of firing shells weighing nearly 18kg (39.68lb).

The Portuguese apparently established a small foundry in Kyūshū. Little is known about this save for a cryptic 1776 Dutch description of a cannon that the Portuguese had manufactured in Japan. Likewise, Japanese copper proved to be of such value that the Portuguese continued to import it to Macao, and the guns they crafted there had the reputation of being the best in all of Asia. One of these Macao weapons was still being used as late as 1812.

FORTIFICATIONS

The existence of powerful cannon led to marked changes in fortification techniques, as stone walls were relied upon and castles came to occupy large enough areas to prevent cannon from bombarding their inner walls. As early as the 1540s weak sections of walls were fortified with large stones. Less than a decade later, impressive stone walls were built for castles, such as Takeda Shingen's Kaizu. Kannonji Castle, a formidable mountain castle, as large as Uesugi Kenshin's Kasuga Castle, was abandoned in 1568 by Rokkaku Yoshiharu (1545–1612), and yet it contained extensive stone walls. These structures were well built, but not particularly high, measuring at the most twice the height of a man. The principle of construction remained that of a mountain castle, in that small mutually supporting structures would cover a larger mountain.

Oda Nobunaga, the destroyer of the Rokkaku, entered the capital in 1568 and in the following year he began construction of Nijō Castle, the abode of the final, and soon to be deposed, Ashikaga shogun Yoshiaki. The Nijō structure was completed in 70 days, and unusually its walls were completely made of stone and it possessed a large moat on the outside, and another on the interior. Louis Frois, when writing about this castle, stated, 'As there was no stone available for the work, [Nobunaga] ordered many stone [Buddhist statues]

Oda Nobunaga used stone to build Nijō Castle for Ashikaga Yoshiaki, the last Ashikaga shogun, in 1568. As good stone was in short supply, he had his troops deface Buddhist statues and used them, an act that horrified many devout Buddhists.

WEAPONS AND FIGHTING TECHNIQUES OF THE SAMURAI WARRIOR

An image of Korean and Chinese troops attacking a Japanese castle (wajō) that had been built in Korea in the 1590s. These castles proved so formidable that none were ever taken, and when the Japanese finally withdrew in 1598, they did so with minimal casualties. The stone walls of these castles are well preserved along the coast of southern Korea to this day.

to be pulled down, and then tied ropes around the neck of these and dragged them to the site.' These walls were 'six or seven ells high [6.86–8m or 22ft 6in–26ft 3in] and six–eight ells [6.86–9.14m or 22ft 6 in–30ft] in width, and in addition three large gates, fortified with stone were also constructed.'

This castle contrasts with one built by Honganji Temple, in the Yamashina suburbs of the capital in 1478, and which was burned and destroyed in 1532. This Honganji fort epitomizes earlier techniques, for it contains two moats and three concentric walls, over 3km (1.86 miles) in length, large enough to protect a small city. Some of the moats, recently excavated, were nearly 4m (13ft) deep, while the walls themselves were 7m (23ft) high. In contrast to Nijō, no stone was used at all.

Not long after Nobunaga constructed Nijō Castle, he commenced building a radically new castle in 1576, in a place called Azuchi. This structure had such extensive stone walls that some scholars have thought that it reveals profound European influences. The Jesuit Luis Frois would praise this structure as one that 'may well be compared with the greatest buildings of Europe' and he marvelled at the stone walls. As the scholar of sixteenth-century Japan George Ellison stated, 'the tower was a colossal structure, which soared 138 feet [42m] into the air form the top of a hill, which itself rose 360 feet [110 m] above the waters on an inlet of Lake Biwa.' Thousands of men were required to push stones up the mountain, with one, known as the 'snake stone' required 10,000 men to move it. Nobunaga also constructed a road some 9m (29ft 6in) wide, guarded by walls 3m (9ft 10in) tall and 3.6m (11ft 10in) wide.

CASTLES AND POWER

Castles were monuments to political power as much as if not more than they were defensive structures. Azuchi, for all of its majesty, had a stairwell rising up the hill, which did not enhance its defences, and the great stone of Ōsaka Castle is merely for show. A visitor to these castles marvels at the elaborate walls, and the soaring keep (*tenshukaku*). Squeaky floor boards and elaborate traps were designed to stop assassins, but militarily if an army could approach to the base of a castle's keep the structure could be easily burned. That is what happened to Azuchi Castle, for Oda Nobunaga's lieutenant, Akechi Mitsuhide, burned it shortly after he killed Nobunaga in 1582.

No castles survive in their entirety, although the structures at Matsumoto and Himeji are the most intact large castles. Matsumoto retains its keep and inner walls, but none of the outer structures remains. Himeji, by contrast, is the best preserved, and retains its inner and secondary wall, but the third wall, which enclosed much of what now constitutes the city of Himeji no longer exists. For a castle to survive a siege, however, the third wall, located kilometres away from the central castle, needed to be preserved, for only by occupying a region of at least 2–3km (1.2–1.9 miles) around the castle could the inner structures be protected from artillery. Some of the largest castles, such

LEFT: *Beginning in the 1560s, Japanese castle walls were constructed by using well shaped stones along the edge, with the sides later filled in. This curved pattern allowed for great strength and stability. Although the corner stones had to be carefully shaped, other stones used for the walls could be either simply stacked or, in the case of castles for the most powerful* daimyō, *carefully fitted together as well.*

BELOW: *Ōsaka castle. Toyotomi Hideyoshi had* daimyō *from all over Japan give him massive stones to be used for his castle at Ōsaka. Its walls were so extensive that his heir Hideyori was able to hold off a besieging army in 1615. Tokugawa trickery, and the filling in of the outer moat after a truce had been called, allowed Tokugawa Ieyasu to batter this castle into submission and destroy the Toyotomi later that year. Only the inner walls and a reconstruction of the keep (*tenshukaku*) are visible here.*

SIEGE TACTICS

The large Japanese castles were vulnerable to fire. Any time that a besieging army could close in to the centre of the castle, it would easily be burned. Here marksmen fire into the castle and archers shoot flaming arrows, while foot soldiers burn stacks of bamboo near its walls. In order to prevent this, many castles had moats. At the same time, some, such as Oda Nobunaga's Azuchi Castle, had a great staircase leading to its central keep, which allowed for it to be burned easily.

as that of Kanazawa, had networks of moats that extended over 15km (9.3 miles) in length.

THE TOKUGAWA AND CANNONS
Although, as we saw in the Introduction, Toyotomi Hideyoshi was an innovator in many ways, he did not rely on cannon heavily as he gained control over Japan in the aftermath of Oda Nobunaga's death. He engaged in a protracted campaign in Korea, and armoured ships the likes of which Nobunaga had created undoubtedly helped him here, although the Koreans devised similar vessels as well. While Hideyoshi invaded Korea, his forces built extensive castles with stone

OPPOSITE: *The Hashizume Gate and Tsukizuki Tower at Kanazawa. Note the moat, and the skill with which stones in the walls were fitted together making this castle one of the strongest in Japan.*

walls, revealing how well the Japanese has mastered these techniques, but he seems to have suffered from insufficient cannon. Surviving edicts demand that all extant cannon be sent to his Nagoya headquarters, and also give orders for the manufacture of many new cannon. Likewise, in spite of the existence of formidable iron ships, Hideyoshi's forces were outclassed on the high seas by similar Korean armoured boats.

After Hideyoshi's death, some cannon were used in the battles of 1600, but the epic battle of Sekigahara, when the supporters of Hideyoshi's son Hideyori were soundly defeated, was decided more by treachery that tactical brilliance or technological prowess. Cannon were, however, used by the besiegers of Fushimi Castle in 1600. Tokugawa Ieyasu, who strove to become the leading warrior of Japan, realized the tactical significance of artillery in reducing

*The keep of Matsumoto Castle, known informally as the 'Raven Castle' (*Karasujō*) from a footbridge leading around its extensive moat. Matsumoto's five-storey keep dates from 1597.*

enemy fortresses. He had to, for Hideyori was ensconced in the greatest castle of all of Japan, located in Ōsaka.

CANNON AND THE SIEGE OF ŌSAKA CASTLE
Immediately after his victory in 1600, and his appointment as shogun in 1603, Tokugawa Ieyasu set out to accumulate as many cannon as possible. He did so, undoubtedly, because he was planning to attack Hideyori, although he did not devise a plausible excuse to do so until 1615. Ieyasu had confiscated nine cannon from the Chōsokabe after the battle of Sekigahara in 1600, and likewise he ordered ten from the Dutch. Although technically not an artillery

INNOVATIONS IN NAVAL WARFARE

The late sixteenth century witnessed, in addition to innovations in castle design, more fleeting improvements in arming ships. Ise craftsmen were, according to the Jesuit Organtino Grecchi-Soldi, able to build a boat in 1578 that 'resembled ships of the Portuguese' and contained, to Organtino's surprise, three cannon. The nature of these cannon is unknown, although it seems most likely that they were breech-loading pieces. According to the *Tamonin nikki,* these boats, which were commissioned by Oda Nobunaga, had a crew of 5000 (!) and were 36m (118ft) long by 21m (69ft) wide and covered with steel plates. This boat effectively blockaded the harbour of a fortified temple, but such craft do not appear to have been used when Toyotomi Hideyoshi invaded Korea in 1592. The ship illustrated is an *atake bune*, a popular style of ship used in the invasion of Korea.

piece, a gun made in the city of Sakai in 1610, with a length of 3m (9ft 10in) and weighing 135kg (298lb), was designed to fire at a distant castle. Ieyasu also had a Sakai craftsman named Shibatsuji Ri'uemon construct a massive cannon that could fire a projectile of 5.6kg (12.35lb). Even larger artillery pieces were constructed as well, with the largest specimens shooting balls of 11.25kg (24.8lb), 13.13kg (28.94lb) and, most impressively, a ball weighing 18.75kg (41.34lb). These guns were extremely short but thick, with the barrel ranging from 7 to 50cm (2.76in to 19.69in). In addition to firing projectiles, these could also fire arrows.

Gunsmiths from Tosa, where the Chōsokabe had resided, were capable of manufacturing guns weighing 60kg (132.28lb) that were capable of shooting a 0.788kg (1.74lb) projectile of 54.5mm

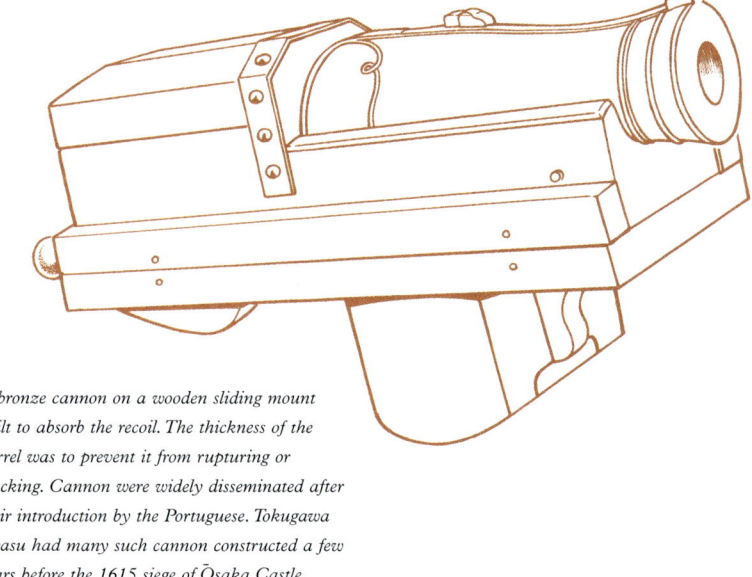

A bronze cannon on a wooden sliding mount built to absorb the recoil. The thickness of the barrel was to prevent it from rupturing or cracking. Cannon were widely disseminated after their introduction by the Portuguese. Tokugawa Ieyasu had many such cannon constructed a few years before the 1615 siege of Ōsaka Castle.

CANNON AND ARTILLERY

A scene from a seventeenth-century illustrated manuscript entitled The Life of Yoshitsune *provides an accurate representation of the outer gate, stone walls and moat of a castle of the time.*

(2.15in) width; another gun could fire a 1.125kg (2.48lb) projectile of 57.6mm (2.27in) for 2.5km (1.55 miles). The cities of Sakai and Nagahama remained important centres of artillery/gun production during the early decades of the seventeenth century. From 1604 until 1615 over 600 weapons can be verified as being constructed by the Kunitomo, with the vast majority of these weapons being guns, and approximately 23 being of large enough bore to be considered cannon. Three of these weapons outclassed others in scale, and all seem to have been constructed to help Tokugawa to defeat the Toyotomi forces in 1615.

Tokugawa Ieyasu utilized the fact that he had been 'cursed' by the inscription on a temple bell – the 'curse' was in fact an innocuous phrase that contained the characters of his name, 'ie' and 'yasu' with a third character inserted between these other two. Ieyasu saw this breaking of his name as a metaphor for the dismemberment of his body and thus a cause for war. This temple bell was for Hōkōji, a temple built to honour Toyotomi Hideyoshi. When the Tokugawa armies first attacked the castle of Toyotomi Hideyoshi's son Hideyori in 1615, they could not batter its walls and their armies could not approach the castle close enough to overwhelm its defences. Their long-range guns were, however, able to damage the keep, and this alone allowed for an armistice to be signed. Some of the Tokugawa gunners, most notably Makino Seibei, Inadome Kunai Shigetsugu and Inoue Geki Masatsugu, proved so effective at bombarding the living quarters of Toyotomi Hideyori and his mother Cha Cha, that according to some accounts, this skill in gunnery forced the Toyotomi's initial surrender.

Once hostilities had ceased, the Tokugawa destroyed the third, outer moat of Ōsaka Castle, thereby making it vulnerable to attack. When the Tokugawa raised a second army a half year later, Ōsaka Castle was easily taken. Through being able to approach the second, rather than the third wall of the castle, it could be more thoroughly pounded by artillery.

THE MONOPOLIZATION OF FIREARMS

Even before the Tokugawa's major rival, the Toyotomi, were destroyed, Tokugawa Ieyasu issued regulations to the major gunsmiths in 1607. After exhorting the smiths to produce many firearms quickly, the Tokugawa then admonished them not to make any weapons for other *daimyō*. Gunsmiths were also prohibited

ABOVE: Detail from a screen of the siege of Ōsaka Castle, 1615. Note how most footsoldiers carry pikes, and the variety of crests depicted on their flags and uniform, which identified them as followers of particular daimyō.

RIGHT: Civilians flee amidst the fall of Ōsaka Castle, 1615. The moat and walls are visible at the top of the illustration.

from travelling to other provinces, and ordered not to inform anyone of the proper recipe for gunpowder, or to teach techniques of gun manufacturing to anyone save for officials of the Tokugawa *bakufu*. In effect, the Tokugawa favoured the more reliable Kunitomo smiths of Nagahama, more than their brethren in Sakai, for the manufacturing of guns, so much so that in contrast to Sakai, the dwellings of the Nagahama smiths have been preserved as a museum. Other smiths, such as Watanabe Sūkaku, who had cast cannon for the Ōtomo, were

CANNON AND ARTILLERY

MISSILE FIRE

At all times, projectile fire formed the mainstay of battle. The sixteenth century witnessed a major shift from bows to guns, which increased the killing range of weapons to 50 yards or so. Forces of skirmishers were protected by pikemen.

ŌSAKA CASTLE, 1615

Three years after his victory at Sekigahara in 1600, Tokugawa Ieyasu was made the shogun of Japan. Toyotomi Hideyoshi's son Hideyori remained a potent force, for he had close ties to the court and was ensconced in Japan's greatest castle at Ōsaka. In 1615, based upon a flimsy pretext, Ieyasu (red forces) besieged the castle, but could not take it. Having declared a truce, his armies filled in the outer moat, and then, six months later, he returned and captured the castle, killing Hideyori in the process.

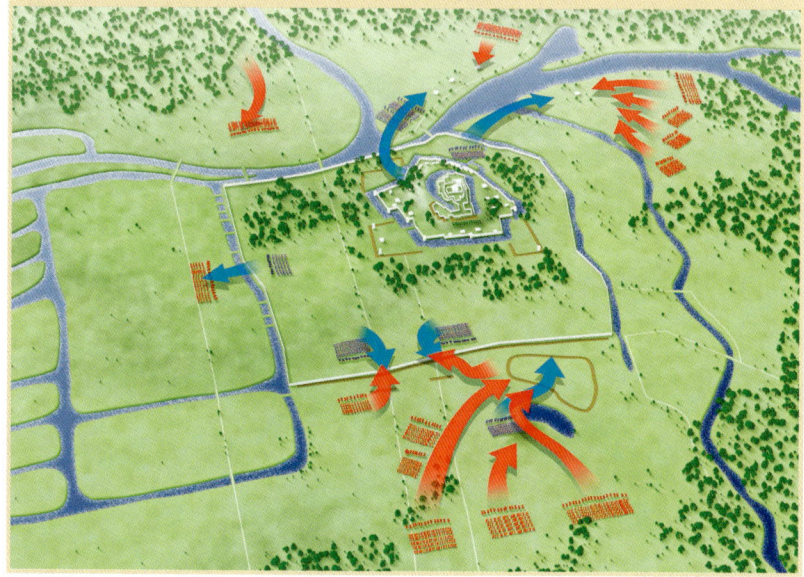

employed by Tokugawa Ieyasu, and manufactured a *furanki* cannon for use in the 1615 campaign. After the battle, Watanabe Sūkaku even collected discarded copper and bronze for use in casting more weapons. The Tokugawa were not alone in crafting these weapons, other *daimyō*, such as the Mōri, a *daimyō* of western Japan, brought 1400 muskets and a few cannon to the Ōsaka encampment in 1615.

Not all of these cannon, particularly the *furanki*, were well made, and one when fired in central Japan caused seven fatalities and 60 injuries when it exploded. Perhaps this is why Tokugawa Ieyasu later relied on purchasing so many from the English or the Dutch. The Dutch ship *Liefde* ran aground in Japan in 1600 and this proved a boon to Ieyasu when it was captured. Its pilot, Will Adams, became a trusted advisor to Ieyasu (and the basis for the fictional hero in James Clavell's novel *Shogun*), and the Tokugawa leader substantially improved his arsenal when he gained control of its 18 cannon, 500 muskets, 5000 bullets, 300 cannon balls and 2268kg (5000lb) of gunpowder.

CANNON AND ARTILLERY

SCALING WALLS
An ingenious method of scaling walls with men climbing over pikes thrust into an enemy's wall. Such attacks could only meet with success when an enemy was surprised or outnumbered. In cases where the wall was manned, these attackers would endure severe casualties.

survives to this day, in no small part because during World War II, when the order went out the melt down iron and bronze objects for munitions, these cannons were buried instead.

TURNING INWARD
A major uprising of Christian converts at Shimabara in 1637 led to the closing of Japan to most trade with Europe. The Shimabara rebels occupied an abandoned castle, located on the coast of Hizen Province and, led by samurai who had followed Christian *daimyō* such as the Konishi and Arima, they fought off the Tokugawa *bakufu*'s army. Cannon from inside the castle killed Itakura Shigemasa, the *bakufu* commander of the siege. Ultimately, the *bakufu* debated whether to use mortars, but the terrain did not favour it. They relied in the end upon the Dutch to bombard the castle from the ocean. After Shimabara fell, and its defenders were slaughtered, the *bakufu* issued a blanket edict prohibiting Christianity, the movement of Japanese abroad or the return of those who had left, and confined a European presence, limited to the Dutch, to a small island in Nagasaki Harbour, called Dejima. These tight controls and other surveillance precluded *daimyō* from procuring weapons from abroad.

The Tokugawa did not confine their interest to guns themselves, for already in 1632 they had established a bureau designed to control the dissemination of bullets and gunpowder. Nevertheless, after the Tokugawa had achieved a position of political hegemony, they deemed it unnecessary to continue to innovate in weaponry. Instead the Tokugawa forced smiths to congregate in easily supervised areas, and they

OPPOSITE: *The bell of Hōkōji, completed in 1614 by Toyotomi's heir Hideyori, contains two passages, highlighted here, that contained double meaning. One stated, 'Rulers and ministers enjoy bounty' could also be interpreted as, 'The people rejoice in Toyotomi rule' while the other, asking for 'Peace in the realm' divided the personal name of 'Ieyasu' in two, which the wily Tokugawa interpreted as a curse. This inscription overruled Ieyasu's oath to support the Toyotomi, which he had signed in 1598, and he attacked and destroyed Hideyoshi's heir in 1615.*

Even after the Ōsaka campaigns of 1615, the Tokugawa continued to import cannon so as to bolster their dominance. Macao became a centre of cannon manufacturing in 1620, and weapons from here were used as late as the eighteenth century in conflicts. In 1624, the Tokugawa imported five cannon from England and 12 from Holland. Not to be outdone, the Shimazu imported at least one cannon from the Portuguese in 1630. This impressive bronze weapon

Shimabara Castle, site of a major Christian rebellion in 1637. The defenders skilfully used their cannon to kill the first Tokugawa commander, but a Dutch blockade and bombardment, and long siege, finally wore the defenders down. The keep and buildings are a modern reconstruction.

prohibited other *daimyō* from manufacturing cannon or learning how to produce these weapons, or for that matter, acquiring the best gunpowder recipes. Some established *daimyō*, such as the Uesugi, managed to maintain a tradition of manufacturing guns and attracted some smiths, such as the Hino, to their domain, now located in Yonezawa. Nevertheless, the Tokugawa had become so powerful that when the Hino smith died in 1619, the Uesugi could not replace him. Within a decade, the Uesugi could no longer manufacture weapons as skillfully as they had done before. The Tokugawa monopolized the

CANNON AND ARTILLERY

manufacture of guns, and slowed the rate of dissemination. Even when they were aware of new inventions, they did not allow this knowledge to spread. For example, we can document flintlocks first appearing in Japan in 1643, and a Dutch captain gave another example to the eighth Tokugawa shogun, Yoshimune, in 1721, but these weapons did not achieve widespread popularity, or use in Japan.

DUTCH KNOWLEDGE
The Tokugawa strove to limit knowledge of important military technologies, and monopolize the access to these weapons. When confronted with the prospect of real innovation in Europe through their limited contacts with the Dutch, the Tokugawa encouraged some samurai to master 'Dutch' knowledge. They established a bureau for the translation of these texts in Nagasaki, the site of where a few Dutch traders were confined. Once the Tokugawa loosened import restrictions on European texts, Western astronomy, medicine and science became known to a small but influential group of samurai who resided in Nagasaki. Secure in its power, the Tokugawa even allowed some peasants the ability to possess guns in 1717, so that they could hunt deer and other animals that interfered with agricultural production.

Some Tokugawa officials, in a private capacity, also began to investigate military technology. In 1832, the 34-year-old Takashima Shunhan, supervisor for the defence of Nagasaki Harbour, developed an interest in cannon and foreign weapons. He was granted permission from the Nagasaki administrators of the Tokugawa regime to study Dutch gunnery, and he did so, purchasing several hundred firearms from the Dutch ranging from muskets to cannon. Five years after Takashima began his interest, two-wheeled cannon were used in Japan in 1837, when the Tokugawa quelled a rebellion by Ōshio Heihachirō, who burned a quarter of the city of Ōsaka in 1837. Only two blasts of projectiles of only 37.5g (1.32oz) in size were necessary to end this uprising. Although these, and other cannon, were useful in crowd control, they were not nearly as effective as the weapons that had bombarded Ōsaka Castle in 1615.

Artillery did not witness dramatic improvements until the mid nineteenth century. Emblematic of the stasis in cannon technology, the best cannon of Macao, which had been crafted early in the seventeenth century, were still used in Europe as late as 1812. There was little improvement in cannon over the course of the eighteenth century, although innovations were made in gun carriages, particularly for use on ships.

199

NINETEENTH-CENTURY CANNON

A cannon like this was used against Ōshio Heihachirō's rebellion of 1837. Heihachirō attempted to redistribute wealth, and attacked large merchant houses and in the ensuing riots, 10,000 dwellings in Ōsaka were destroyed. Three cannon were wheeled into the streets of Ōsaka to quell the riot and they were effective in controlling the crowds, although these cannon were not of particularly high quality construction.

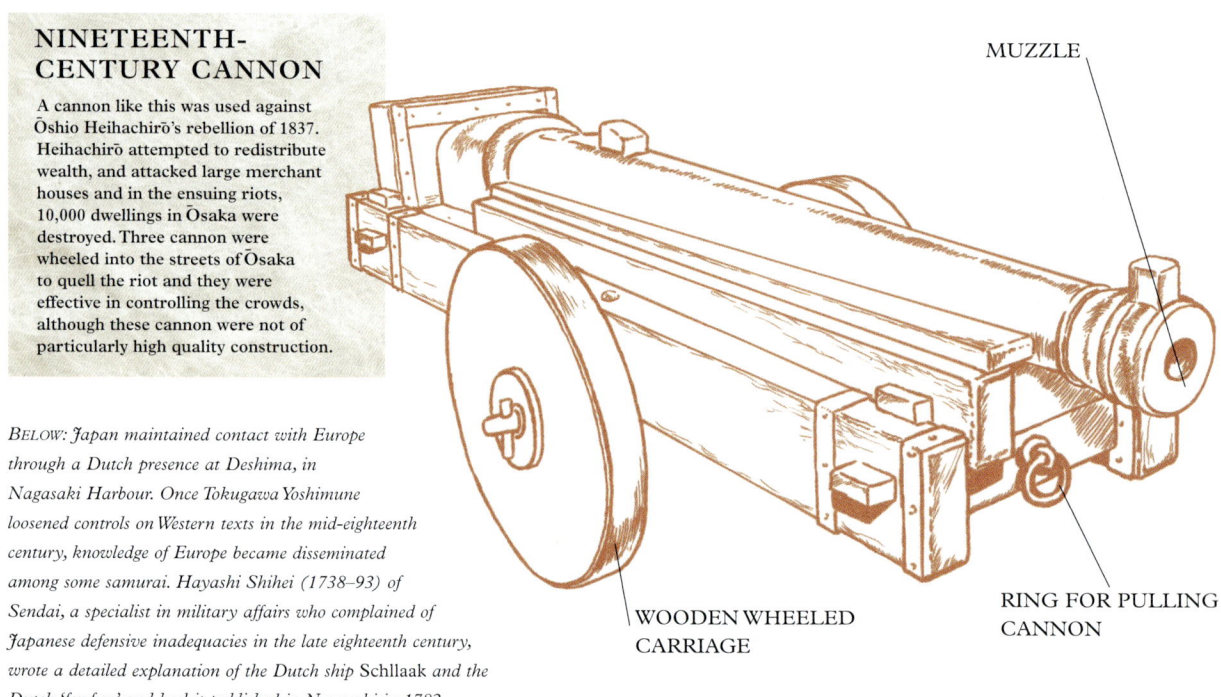

MUZZLE

WOODEN WHEELED CARRIAGE

RING FOR PULLING CANNON

BELOW: *Japan maintained contact with Europe through a Dutch presence at Deshima, in Nagasaki Harbour. Once Tokugawa Yoshimune loosened controls on Western texts in the mid-eighteenth century, knowledge of Europe became disseminated among some samurai. Hayashi Shihei (1738–93) of Sendai, a specialist in military affairs who complained of Japanese defensive inadequacies in the late eighteenth century, wrote a detailed explanation of the Dutch ship* Schllaak *and the Dutch 'for fun' and had it published in Nagasaki in 1782.*

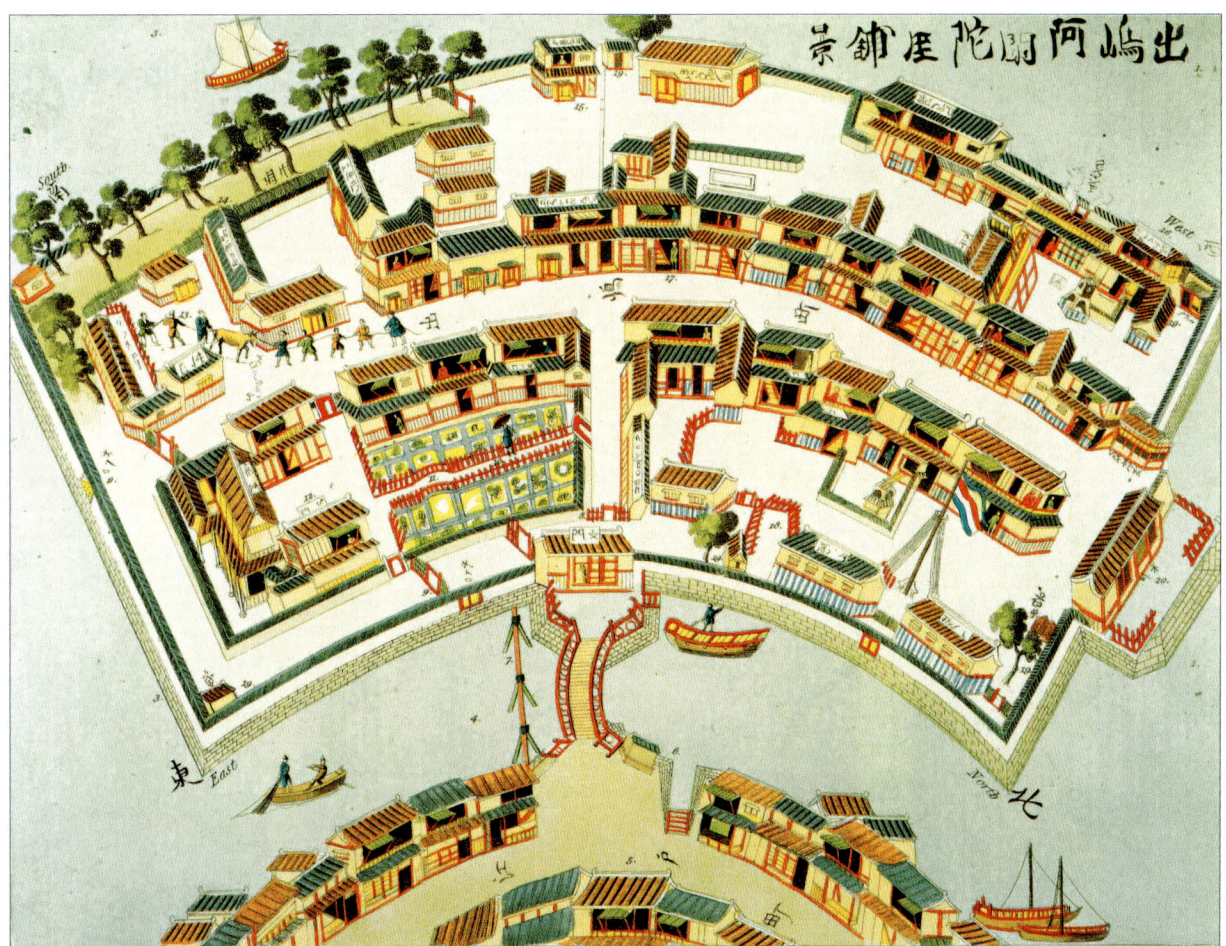

Not until the 1840s do improvements in cannon again become pronounced, and indeed it was at this time that the rifling of cannon barrels began. This period of innovation was coupled with a geopolitical shock, for the British defeated the Chinese during the Opium War. This event resulted in increased interest in new weapons. Takashima, in particular, issued a blistering critique of Japanese armaments, stating that they were centuries out of date. He even paraded a small unit using Western drill in 1841. Interest in and attempts to manufacture new, improved weapons increased at a dramatic pace.

Samurai involved with Dutch studies mastered the theoretical underpinnings of industrial production, with their greatest interest focused on armaments. One such samurai, Fukuzawa Yukichi, learned how to tin plate iron, and how to manufacture chloric and sulphuric acids, and even had access to state-of-the-art books about electricity. Fukuzawa himself so identified with the new ways of acting that he sold his swords; clearly, for many the new ways of technology mattered more than the ideals of the past.

The province of Saga built the first reverbaratory furnace in 1850, using a Dutch book as a guide. An iron gun was satisfactorily cast from one of these furnaces in 1853. Over 100 workers were employed at this factory, which completed an order for 200 cannon by 1857. In 1852, samurai from Saga domain studied photography, the telegraph and steamships, and were able

The Dutch were confined to this small island in Nagasaki Harbour in 1641, and only allowed to leave once a year on a procession to visit the Tokugawa shogun.

to make models of the last two before Commodore Perry arrived.

This knowledge rapidly spread to other provinces. Men from Satsuma, located on the periphery of Japan, were exposed to considerable foreign knowledge through their interactions with the Ryūkyū islands, and Shimazu Nariakira (1809–58), the *daimyō* of Satsuma, even wrote his diary in Roman letters so as to keep his actions private from the *bakufu*. In 1851, Shimazu Nariakira had created the Seirenjo, which allowed him to smelt iron ore and by 1858 the domains of Satsuma and

ŌYOROI ARMOUR

NINETEENTH CENTURY, LATE EDO PERIOD

Shimazu Nariakira (1809–58) was an innovative *daimyō* of Satsuma who was interested in European manufacturing and technology. He even wrote Japanese in his diary with roman letters, so as to confound the Tokugawa censors. In contrast to this openness to new ideas, Nariakira's armour represents a revival of the old *ōyoroi* style of armour. Save for the face and neck protection, and the *kote* gloves, his suit is an accurate reproduction of thirteenth-century armour. The bearskin shoes (*tsuranuki*) he wears had fallen from favour half of a millennium previously.

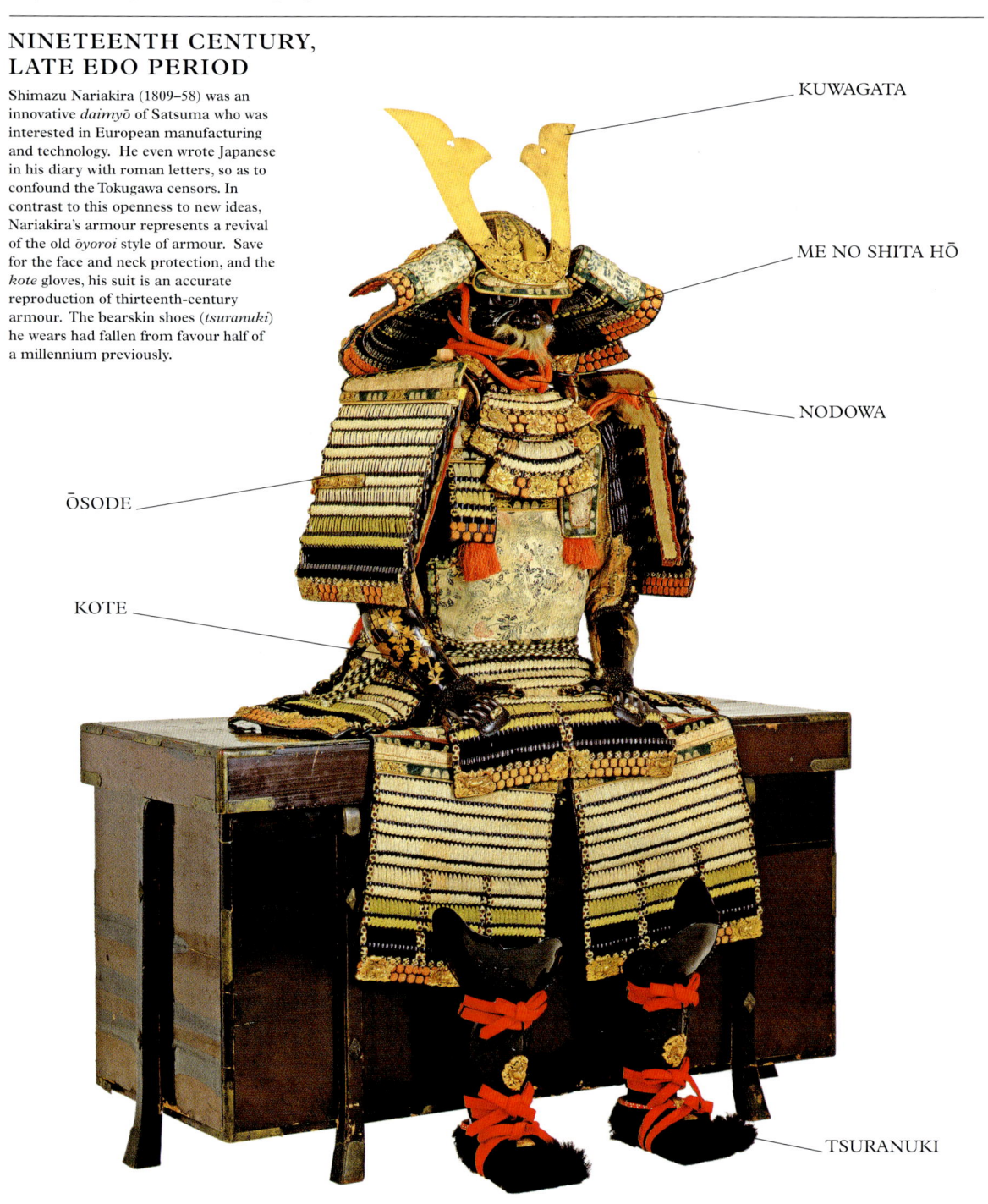

CANNON AND ARTILLERY

LEFT: Fukuzawa Yukichi, a proponent of 'Dutch Studies', thought so little of his swords that he sold them and replaced the blades with bamboo. He went on several missions to Europe and America and became a major thinker and proponent of 'civilization and enlightenment' in the Meiji era (1868–1912). He is depicted here on the far right of this 1860 photograph.

Mito, and for that matter, the Tokugawa themselves, had built blast furnaces. Satsuma gained the ability to rifle cannon bores in 1865, as sketches from a Dutch book allowed for the construction of an apparatus for boring solid cast-iron gun barrels from furnaces.

A series of visits by Commodore Perry of the United States caused the Tokugawa to finally open Japan to more direct contact with America and the nations of Europe in 1854. Perry and his men displayed small steam engines and the telegraph so as to impress the 'feudal' Japanese, even though practitioners of Dutch studies already had a theoretical understanding of these objects. Indeed, smiths in Satsuma had built three model steamships in 1852, the year before Perry visited Japan, and by 1855, they had constructed a small steamship. That same year, the Tokugawa built their first Western-style ship.

At the same time as some samurai were immersed in Western technology, others developed a sense of Japanese superiority. Proponents of this school of

BELOW: Uesugi Shrine, Yonezawa Yamagata, Japan. The Uesugi fought against the Tokugawa in 1600 and had most of their wealth confiscated and their lands transferred to the Yonezawa.

The opening of Japanese treaty ports in 1858, four years after Commodore Perry first forced the Tokugawa to end their policy of seclusion against the west, allowed haughty westerners to come into contact with equally proud samurai. Assassinations and random attacks became common in the early 1860s.

thought, known as Nativism, as well as those specialists in Dutch learning, tended to congregate in the same western provinces. Satsuma, far to the south, epitomized these contradictions.

A sense of pride in the samurai order and its long history pervaded Satsuma, and yet also interest in blast furnaces and manufacturing cannons. Anti-foreign sentiment became strong after Japan's opening, and when a British merchant named Richardson rode in front of the retinue of the Satsuma *daimyō* and was killed. Britain demanded an exorbitant indemnity of 125,000 British pounds (100,000 from the Tokugawa and 25,000 from Satsuma), which in purchasing power today would be the equivalent of 20 million dollars (10 million British pounds). The British later bombarded Kagoshima, the main city in Satsuma domain, in May of 1863, and the

Satsuma gunners concentrated their fire on the British flagship, killing the British commander-in-chief and captain. Kagoshima suffered grave damage, but the British were driven into fits of paranoia, for they thought it impossible that the Japanese could manufacture an explosive shell and assumed instead that the Japanese were aided by the Russians.

Confronted by rebellious provinces, and struggling to have their foreign treaties accepted by the court, where sovereign authority remained, the Tokugawa loosened restrictions on *daimyō* in 1863, and this allowed them to

CANNON AND ARTILLERY

Ultimately, the resurgent western domains attacked and defeated the Tokugawa forces, and Tokugawa Yoshinobu (1837–1913) resigned his position as shogun in 1867. Once the Tokugawa were overthrown, the Satsuma and Chōshū leaders had the emperor issue a charter oath, which stated that knowledge was to be sought throughout the world. This marks the beginning of the Meiji era (1868–1912), that was meant to be an age of 'enlightened rule.'

THE END OF THE SAMURAI

The Meiji state rapidly became more centralized. 'Evil customs' of the past were abandoned in favour of 'accepted practices throughout the world', which led to a profound questioning of Japanese identity. Increasingly, the samurai and their swords were perceived to be an element of the past. By 1872, the Meiji government had secured direct control over all *daimyō* domains, and abolished the samurai order, converting samurai stipends to bonds. At this time they gained control over the weapons

A nineteenth-century cannon, manufactured by the Takashima school, revealing the skill with which the Japanese could cast cannons even before Japan was 'opened' by Perry in 1854.

congregate in the old capital and purchase munitions from abroad. Plotting continued, until an alliance of domains, including Satsuma and Chōshū (two who had lost to Tokugawa Ieyasu at Sekigahara in 1600), attacked and overthrew the Tokugawa *bakufu* in 1868. Their clever slogan, 'revere the emperor and expel the barbarian' alluded to the weakness of the Tokugawa shogun in resisting the 'barbarians' of the West. This phrase appealed to Japanese chauvinists and legitimized their insubordination by professing allegiance to a powerless emperor.

Sakurajima, the very active volcano that overlooks Kagoshima Harbour. This photo was taken from the gardens of the Shimazu, the daimyō *of Satsuma domain.*

Tokugawa Yoshinobu, Japan's last shogun. He became shogun too late to successfully institute his far-reaching reforms.

customs of 17,000km (10,000 miles) away. A few leaders of the 1868 overthrow of the Tokugawa felt dissatisfied with the rapid pace of change.

Portrayed as reactionaries, many men longed for the re-establishment of the samurai order, although they also had a disdain for what they perceived as despotism of the Meiji state. Eto Shinpei (1834–74), who had translated Napoleonic Codes into Japanese, led an uprising in 1874, which was crushed, as too was another in Chōshū the following year. Another disgruntled leader, Saigō Takamori (1827–77) from Satsuma, led 15,000 men on an attack of Kumamoto

each domain had, and in the end confiscated over 180,000 guns, mostly Enfield or Stahl muzzle-loaders that had been imported after 1863, when the *bakufu* had lifted most of its restrictions on *daimyō* and military trade.

The abolishment of the samurai order left many adrift. *Daimyō* and high-ranking samurai had enough wealth, in the form of bonds, to have a comfortable existence, but others needed to find new professions. Some chose the army, while others strove to be educators, officials, editors or even barbers (for they were skilled with a sharp blade).

Tensions remained, particular in the triumphant areas of western Japan, which had overthrown the Tokugawa. Centralization meant that old geographic names and identities were lost. Some in Chōshū would resist, for example, wondering why they should follow the

CANNON AND ARTILLERY

Castle. His men failed to take it after a 55-day siege. The army suffered casualties of 6000 out of 65,000, while in the end nearly all of the rebels were killed, wounded or captured.

REBEL INSURRECTION

Saigō Takamori and his men relied on mostly muzzle-loaders, such as Geballe and Enfield rifles, and suffered particularly in the dampness of Kyūshū because their powder could get wet, and therefore rendering the weapons inoperable. By contrast, the government forces used Snider rifles, which were breech-loading and did not suffer from their powder getting wet. Takamori's rebels also laboured under the difficulty that they possessed a variety of guns of varying calibres, and thus could not procure adequate ammunition – they had only 100 rounds per man. Takamori also had a modest force of artillery: 28 mountain guns firing 2.39kg (5.28lb) shells and two field guns firing 7.18kg (15.84lb) shells and some 30 mortars. By contrast, the newly formed Japanese conscript army had 100 artillery pieces, two Gatling guns and 63 million rounds of ammunition, more than 14 times the amount of the rebel army, and by March of 1878, the government could produce 500,000 rounds per day.

In the fiercest encounter of this campaign, Meiji armies expended 300,000 rounds per day when attacking Takamori's forces. The rebels were unable to take Kumamoto castle with their limited artillery and, lacking adequate ammunition and gunpowder, they finally abandoned their guns for swords. This final desperation has been remembered and depicted in films, but in fact guns had already been part of the samurai method of fighting.

Proud samurai could no longer compete on the battlefield unless they were part of a centralized state, one that was supported by an extensive industrial base. It is telling that 1878 marks the final year of the samurai revolts, and the

The Rebel Insurrection at Kagoshima, by Tsukioka Yoshitoshi, depicts the revered hero of the Meiji Restoration, Saigō Takamori. He retired to Kagoshima province, which had formerly been the Satsuma domain, in 1875, and launched the rebellion against the government he helped to establish in 1877. Here, his samurai army was eventually defeated by the government's modern conscript army. In this artwork an idealized Takamori rallies his troops against a naval attack.

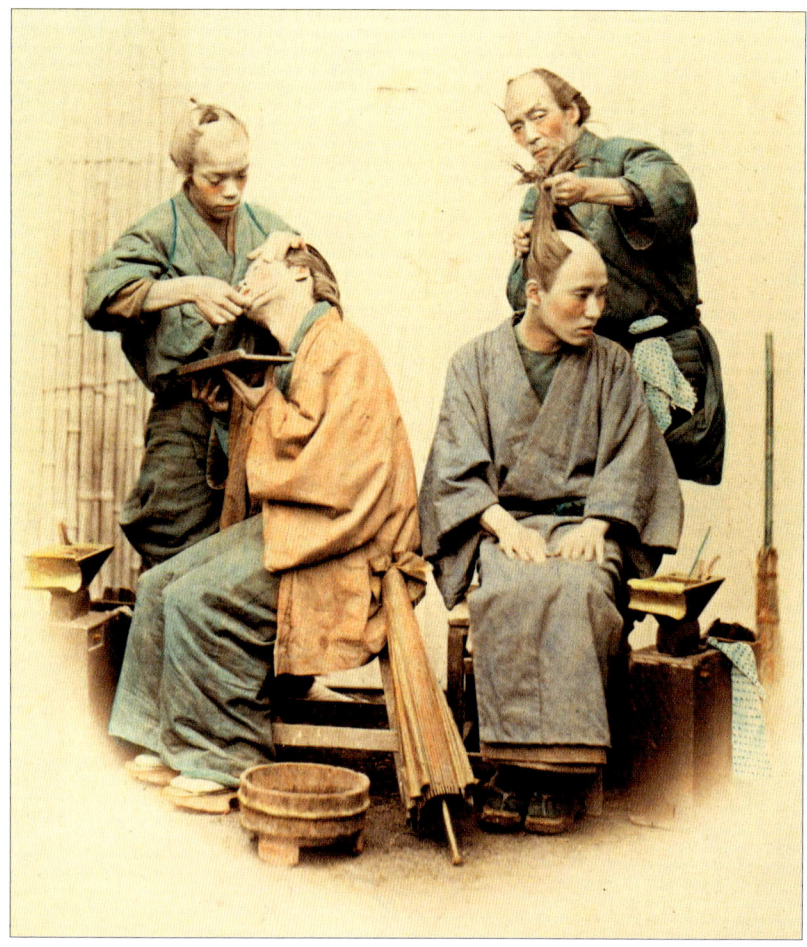

Barbers photographed in 1865 shaving and dressing the hair of customers. When the samurai order was abolished in the early 1870s, some ex-samurai decided to become barbers for it allowed them to continue wielding a sharp blade.

samurai, that their order was founded in death. Interest in gunpowder weapons continued, but the Tokugawa hegemony prevented this technology from rapid dissemination, as it was feared that these guns would threaten the regime.

The defeat of the Chinese by the British spurred interest in guns and cannon, and the Tokugawa loosened controls, allowing domains, particularly western ones, to manufacture guns and final year that the old rifle makers, such as the Kunitomo, ceased making their beautifully crafted harquebuses as well.

CONCLUSION

The fighting techniques of the samurai changed remarkably over time, with first a small band of horse riders, relying on bows and arrows then a shift towards massed tactics and forces, which relied heavily on pikes, and an expansion of the number of those who fought. Accompanying these changes, older more distant patterns of leadership gave way to more intensive, charismatic command. Furthermore, armies expanded from consisting of small bands of horsemen to a large percentage of the able-bodied men in each province.

The adoption of guns proved significant, as these weapons supplanted bows, and led to high casualties in commanders. At the same time, the use of cannon contributed to the creation of massive stone castles occupying plains, and these came to symbolize the new Japan. Toyotomi Hideyoshi created a samurai order, where those who were samurai had to abandon their lands and keep their swords as symbols of their status. Swords too became a symbol of their order, and many thinkers emphasized the importance of martial arts, or the idea, new really to the cannon; study of telegraphy and electricity increased as well, so that even as Japan 'opened' to Perry, his objects, such as the telegraph and steam train, meant to overawe, were already understood. The arming of the domains contributed to a war in 1868 that led to the overthrow of the Tokugawa, and the new government ultimately abolished the samurai order and the old ways of the past. Rebellions ensued, with the last, by Saigō Takamori, fought in defence of samurai privileges.

They failed, however, but their reliance on guns to fight for all but the last desperate moments reveals that the gun and cannon were essential for both the samurai and their rivals in the conscripted army. In the end, the final fighting techniques of the samurai seamlessly melded into the practices of the modern army.

Saigō Takamori's rebellion. This print unusually shows Saigō's forces using guns in the upper right, but as should be evident, his army, upon running out of bullets, was forced to resort to using swords as well. This print over emphasizes the use of swords in battle, particularly by the imperial army, as guns represented the dominant weapon for both sides.

GLOSSARY

ashigaru
Foot soldiers, or skirmishers, most often men who fought on foot.

bakufu
Literally meaning 'tent government', this term describes three warrior governments of Japan: the Kamakura (1185–1333), Muromachi or Ashikaga (1338–1573) and Tokugawa (1603–1867) regimes. A shogun, or 'barbarian-subduing generalissimo', headed each *bakufu*.

buntori
The capture of a piece of an enemy, be it part of the body of a deceased warrior or a piece of his armour.

eboshi
A cloth cap worn by adult warriors that also served as a helmet liner for comfort.

dōmaru
A simplified suit of armour, fasted in the back, but later thought to be fastened on the side. First appeared in the thirteenth century. Suits of *dōmaru* were embellished with accoutrements, including leggings, gloves, shin guards and face guards (*hōate*).

ebira
A basket-like quiver, designed to be attached to a warrior's back.

furanki
Breech-loading cannon from Europe. Also known as *ishibiya*.

gekokujō
A term describing the lower conquering the higher, which was often used to describe the 'Warring States' turmoil of the sixteenth century.

genin
'The base' or people lacking surnames who were generally menial servants, possessing almost no legal rights.

gokenin
'Honourable housemen' – denoted autonomous warriors who performed guard duty for the Kamakura regime. No iron-clad method existed of determining who was in fact a *gokenin*. This term continued to be used after the fall of the Kamakura *bakufu*, but it generally became replaced by the term *tozama*.

gunchūjō
Petition for rewards, documents submitted by warriors demanding compensation for their military service.

gunpai uchiwa
Flat fans used by generals in battle. A symbol of office that became common in the sixteenth century.

gunsen
Ribbed fans used by generals.

haidate
Flexible leg protection.

hanzei
Reference to a law passed in 1351, whereby half of a province's revenue was earmarked for the procurement of provisions and military supplies.

hara-ate
The simplest type of armour that merely protected the chest and stomach.

haramaki
A type of armour fastened in the side, which represented a simplified version of 'great armour.'

hatsumuri
Gear that covered the forehead and the sides of the face, but not the jaw and neck, that was commonly used through the thirteenth century.

hiya
Primitive three-barrelled guns.

hō-ate
Face guards designed to protect the neck and lower face from arrow wounds, which were adopted as a result of the endemic warfare of the fourteenth century.

horo
A cape designed to protect warriors.

hyakushō
Meaning quite literally, '[members of] the hundred surnames', this term initially designated all members of society who held an obligation to pay taxes. The first reference to *hyakushō* as

designating cultivators appears in the late thirteenth century, but even in the mid fifteenth century there were still powerful provincials with surnames referring to themselves as *hyakushō*.

ichimaibari uchidashidō
Armour of the sixteenth century made of two large plates of metal, one protecting the torso and the other the back.

ikki
A verb originally meaning 'to be in accord,' became used as a noun to designate military units during the fourteenth century.

ishibiya
Early breech-loading cannon. *Furanki*.

jitō
A land steward, or one who was invested with the office of *jitō* in the aftermath of the Genpei Wars (1180–85) by Minamoto Yoritomo. Possessing powers of taxation and policing, this post proved most desirable for provincials, for they could only be dismissed by orders of the Kamakura *bakufu*. This office continued to be granted throughout the fourteenth and fifteenth centuries, but it became amalgamated with the social category of *gokenin* during the fourteenth century.

Jōei Code (Shikimoku)
A law code drawn up by members of the Kamakura *bakufu*, which provided great protection for land rights.

kaburaya
A turnip-shaped arrow that emitted an eerie low humming sound, often fired during the onset of battle.

kabuto
A steel helmet.

karimata
Two pronged arrowheads, often used for hunting.

kassen bugyō
Battle administrators who inspected battle reports (see *kassen chūmon*) and wounds in order to determine their veracity.

kassen chūmon
Battle reports, representing a detailed record of wounds that were submitted to battle administrators (*kassen bugyō*), who rigorously inspected the documents, sometimes adding notations making them more precise. These reports often later became summarized in petitions for reward (*gunchūjō*) and hence were less likely to be preserved.

kan
3.75kg (8.3lb) of copper, the equivalent of one thousand mon of cash (Each mon constituted 3.75g (0.13oz). Worth approximately US$1000/£500 today.

kanjō
Written commendations or documents of praise written by commanders to their followers from 1333 onward.

kanrei
The chief of staff of the Muromachi *bakufu* from the time of Ashikaga Yoshimitsu's rule during the late fourteenth century. This post should be conceived as a deputy shogun, and became the preserve of the Hosokawa, Hatakeyama and Shiba families.

Kantō
Eastern Japan, most specifically the eight provinces surrounding Kamakura. During the Ashikaga regime, it was headed by a member of this family who was not the shogun.

Kantō kanrei
The deputy of the east; denotes the Uesugi, who helped the Ashikaga leader of the east.

katana
A curved sword exceeded 0.6m (2ft) in length. Varied slightly from a *tachi* sword. Became a marker of *samurai* status in the sixteenth century.

kebiki
'Hair pulling' – an earlier style of braiding for armour that became obsolete over the course of the fifteenth and sixteenth centuries.

koshitate
Shields, made from wooden boards or doors, that were placed in front of foot soldiers to protect them from arrows. At times, they could be held by charging troops. Shields were often plundered from temples or other dwellings (*zaike*).

kote
Chain metal cloves designed to protect the upper hands and arms of the samurai.

kubō 'Mr Public.'
A term originally used for the emperor or court. During the fourteenth century this word was used designate members of the Ashikaga shogunate, or, on occasion, their plenipotentiaries as individual manifestations of a more public authority.

kumade
Claw-like hooks designed to pull *samurai* off their horses.

kuni kuzushi
Literally 'destroyer of countries', a colourful name for cannon.

kura
A saddle.

kuwagata
Two horns located on the helmet of high-ranking warriors. These were thought to provide protection for its wearer.

maewa
A pommel board.

masakari
Battle axes.

myōshu
Literally 'lord of the *myō*' – described autonomous warriors who were not gokenin.

naginata
A weapon with a long curved blade attached to a wooden staff.

Nanbandō
Japanese armour that used fragments of European armour, most commonly the breast plate, combined with Japanese cloves. Somewhat resembled *ichimaibari uchidashidō*.

nobushi
Also known as *nobuseri*, this word refers to skirmishers of all social rank, and not as a 'class' of foot soldiers, as been sometimes assumed. These skirmishers were generally archers, and were indicative of the nature of fighting during the fourteenth century, when loosely scattered groups of soldiers predominated.

ōdachi
A long sword, otherwise known as a *nodachi*, or 'field sword,' over 0.9m (3ft) in length, which appearing first in the thirteenth century and gained great popularity in the fourteenth. Increasing to 2.1m (7ft) in length, these swords were ideally suited for the scattered battles of the fourteenth century, but were replaced by pikes in the fifteenth and sixteenth centuries as tactics shifted towards massed groups of infantrymen.

ōyoroi
Initially known only as *yoroi*, this term refers to the most expensive armour, which was worn with shoulder boards (*sode*), helmets and all other accoutrements. By far the most expensive armour, *ōyoroi* was made for generals or other high-ranking individuals. This box-like armour was best suited for use on horseback and most effective in protecting against arrows.

samurai
Originally the dependent followers of *gokenin*, this became a general term for all warriors in the seventeenth century.

shōgun
An eighth-century office, originally designed to quell barbarians, and hence known as *sei-i-taishōgun*, this post became important as the highest authority within the bakufu, which contained delegated powers of military and judicial authority. Minamoto Yoritomo was appointed to this post in 1192, but he does not appear to have emphasized this office. During the late thirteenth and early fourteenth centuries, the post of shogun was reserved for court nobles, or princes of imperial blood. With the downfall of the Kamakura bakufu in 1333, Ashikaga Takauji started laying claim to the post again, and was appointed shogun in 1338 (the Ashikaga, or Muromachi *bakufu*). In 1603, Tokugawa Ieyasu received this position, thereby legitimating his family's dominance, which lasted until 1868.

shugo
In the Kamakura era, a 'constable' responsible for investigating crimes, suppressing rebellion and organizing *gokenin* for guard duty. This position, originally held by warriors lacking particular local power, became the focus of ambitious warriors after the promulgation of the hanzei edicts. Prior to 1351, military powers accrued more to generals (*taishō*) but in later half of the fourteenth century *shugo* became incipient local magnates because they could directly extract economic surpluses for the waging of war.

sode
Armour shoulder boards, covering the upper arms of warriors and used to protect the sides of warriors from arrows as they fought on horseback. *Sode*, best thought of as attachable shields, continued to be used only as long as warriors fought on horseback.

sōryō
Family chieftain, one who has the ability to lead other dependent relatives of his family in battle.

soya
These 'war arrows' were designed to penetrate armour and had a thin steel tip.

sugake
An x-shaped weave that was preferred in later styles of armour for it used less braiding.

sune-ate
Shin guards.

tachi
A sword. During the fourteenth century long-swords and, to a lesser extent, naginata were the preferred weapons for hand-to-hand combat. Nevertheless, the bow remained the dominant weapon throughout the fourteenth century. Varied slightly from the more curved blade of a *katana*.

taishō
A general, appointed from the ranks of noted families on a provisional basis. With the fall of the Kamakura *bakufu*, most generals were either collaterals of the Ashikaga lineage, imperial princes or court nobles. Their powers were formidable, but rarely institutionalized. Although *taishō* proved more capable of mobilizing troops than *shugo*, after the hanzei edicts of 1351 *shugo* could more effectively wield military force because they could systematically provision their armies.

tatehagidō
Armour made from vertical plates of armour welded together.

tantō
Short swords and daggers, under 30cm (1ft) in length.

tebiya
Early generic term for 'hand guns', which included harquebuses and more primitive weapons (*hiya*).

teppō
A word that initially described explosive projectiles, but later came to be a common term for guns, particularly Portuguese harquebuses.

tōsei gusoku
Simplified armour with minimal weaving that became common in the sixteenth century.

tōya
Arrows fired at a long distance.

tozama
Literally meaning an 'outsider,' this term refers to warriors who were capable of behaving with autonomy, receiving their own petitions for reward, and fighting for whomever they pleased.

utsubo
A fur-covered quiver. Replaced *ebira* in the fourteenth century.

waidate
A separate piece that fits in the side of 'great armour', thereby making it roomier.

wakatō
Retainers possessing surname, but not autonomy. They were capable of riding horses, but not submitting their own petitions.

wakizashi
A short curved sword, generally from 0.3–0.6m (1–2ft) in length.

yari
Pikes.

yukinoshita dō
Simplified metal armour used by soldiers of the Latter Hōjō. Made for their exclusive use by the Yuki no shita smiths, who resided in the *Kantō*.

yumiya
Literally 'bow and arrow,' but refers to the warrior social order.

yoroi
See *ōyoroi*.

yokohagidō
Metal armor made from horizontal pieces of metal welded together.

zunari
Sixteenth-century helmets, designed to closely fit the head. Many had elaborate decorations.

BIBLIOGRAPHY

PRIMARY SOURCES AND REFERENCE MATERIALS IN JAPANESE

Place of publication, unless specified, is Tokyo.

Aichi kenshi shiryō hen, vol. 11. Compiled by Aichi ken hensan iinkai. Aichi, 2003.

Azuma kagami. Edited by Kuroita Katsumi. In *Shintei zōho kokushi taikei*. 4 vols. Yoshikawa kōbunkan, 1975–77.

Baishōron. Edited by Yashiro Kazuo and Kami Hiroshi. In *Shinsen Nihon koten bunko*. Gendai shichōsha, 1975.

Buke myōmokushō. Compiled by Hanawa Hokiichi. 8 vols. Meiji tosho shuppan, 1954.

Chūsei hōsei shiryōshū. Edited by Sato Shin'ichi and Ikeuchi Yoshisuke. 7 vols. Iwanami shoten, 1955–2005.

Dai Nihon komonjo iewake, no. 9 Kikkawa ke monjo. Compiled by Tōkyō daigaku shiryō hensanjo. 3 vols. 1925–32.

Dai Nihon komonjo iewake, no. 11 Kobayakawa ke monjo. Compiled by Tōkyō daigaku shiryōhen sanjo. 2 vols. 1917–18.

Dai Nihon komonjo iewake, no. 8 Mōri ke monjo. Compiled by Tōkyō daigaku shiryōhen sanjo. 4 vols. 1920–24.

Dai Nihon komonjo iewake, no. 12 Uesugi ke monjo. Compiled by Tōkyō daigaku shiryōhen sanjo. 3 vols. 1931–63.

Dai Nihon shiryō, series no. 6, 8, 12. Compiled by Tōkyō daigaku shiryōhen sanjo. Tōkyō daigaku shuppan, 1901–.

Hekisan nichiroku, in *Zōho zoku shiryō taisei*, vol. 20. Kyōto, Rinsen shoten, 1982.

Inryōken nichiroku. in *Dainihon bukkyō zensho*, no. 133–37. Compiled by Bussho kankōkai hensan. 5 vols. 1912–13.

Kamakura ibun. Compiled by Takeuchi Rizō. 51 vols. Tōkyōdō shuppan, 1971–97.

Kanagawa kenshi, shiryo hen, kodai-chūsei 3.3. Compiled by Kanagawa ken. Yokohama, 1979.

Kawanaka kassenzu Nagashino kassenzu. Edited by Kuwata Tadachika. In *Sengoku kassen-e byōbu shūsei*, vol. 5. Chūō Kōronsha, 1988.

Kokushi daijiten. 15 vols. Yoshikawa kōbunkan, 1979–97.

Kyōto gekidō no chūsei. Kyōto bunka hakubutsukan, 1996.

Nanbokuchō ibun Chūgoku, Shikoku hen. Compiled by Matsuoka Hisato. 6 vols. Tōkyōdō shuppan, 1987–95.

Nanbokuchō ibun Kyūshū hen. Compiled by Seno Sei'ichirō. 7 vols. Tōkyōdō shuppan, 1980–92.

Niigata kenshi shiryō hen chūsei. Compiled by Niigata kenshi hensan iinkai. 3 vols. Niigata, 1981–83.

Sengoku ibun Go-Hōjō hen. Compiled by Sugiyama Hiroshi. 6 vols. Tōkyōdō shuppan, 1989–95.

Sengoku ibun Takeda hen. Compiled by Shibatsuji Shunroku. 5 vols. Tōkyōdō shuppan, 2002–.

Shinchō kōki. Edited by Okuno Takahiro. Kadokawa bunko, 1969.

Shizuoka kenshi shiryōhen chūsei. Compiled by Shizuoka ken. 4 vols. 1989–96.

Taiheiki (Jingū chōkōkan hon). Edited by Hasegawa Tadashi, et al. Ōsaka, Izumi shoin, 1994.

Uesugi-ke gosho shūsei, in *Jōetsu shishi sōsho* no. 6, vol. 1. Jōetsu shi, 2001.

Wachi chōshi shiryōshū chūsei kinsei no. 1. Compiled by Wachi Chō. Wachi, 1987.

Wakayama kenshi chūsei shiryō, vol. 2. Compiled by Wakayama ken. Wakayama, 1973.

Yusoku kojitsu daijiten. Compiled by Suzuki Keizō. Yoshikawa kōbunkan, 1995.

Zōho teisei hennen Ōtomo shiryō. Compiled by Takita Manabu. 33 Vols. Ōita, 1962–71.

SECONDARY SOURCES IN JAPANESE

Fujimoto Masayuki. *Yoroi o matō hitobito*. Yoshikawa kōbunkan, 2000.

Fujimoto Masayuki. *Nobunaga no Sengoku gunjigaku*. JICC shuppan, 1993.

Hamada Toshiyasu. 'Teppō denrai no keii ni tsuite ni san kōsatsu', *Reimeikan chōsa kenkyū hōkoku*, no. 14. Kagoshima ken rekishi shiryō sentaa Reimeikan, 2001: 85–100.

Hayashida Shigeyuki. *Nihon zairaba no keitō ni kansuru kenkyū*. Nihon chūō keibakai, 1978.

Hora Tomio. *Teppō-denrai to sono eikyō*. Kyōto: Shibunkaku shuppan, 1991.

Imatani Akira. *Sengoku jidai no kizoku*. Kōdansha gakujutsu bunko, 2002.

Ishii Susumu. *Chūsei no bushidan*, in *Nihon no rekishi*, vol. 12. Shōgakkan, 1974.

Isogai Masayoshi et al. (eds). *Zusetsu Takeda Shingen*. Kawade shobō, 1987.

Kannonji jō to Sasaki Rokkaku. Azuchijō kōko hakubutsukan, 1995.

Kawazoe Shōji. *Imagawa Ryōshun*. Yoshikawa kōbunkan, 1964.

Kobayashi Yoshiharu, comp. *Nagashino shidaragahara no tatakai*. Yoshikawa kōbunkan, 2003.

Kondō Yoshikazu. *Chūseiteki bugu no seiritsu to bushi*. Yoshikawa kōbunkan, 2000.

Murata Shūzō. *Zusetsu chūsei jōkaku jiten*. 3 vols. Shinjinbutsu Ōraisha, 1987.

Nihon shi ni miru teppō ten. Ōsaka: Yomiuri shinbun, 1972.

Ogawa Makoto. *Yamana Sōzen to Hosokawa Katsumoto*. Shinjinbutsu Ōraisha, 1994.

Ōmijima no dōmaru. Compiled by Ōyamazumi jinja. Ōyamazumi shrine, Ehime, 1991.

Ōmijima no haramaki. Compiled by Ōyamazumi jinja. Ōyamazumi shrine, Ehime, 1991.

Owada Tetsuo. *Hōjō Sōun to sono shison*. Seibunsha, 1990.

Sakai Naoyuki (ed). *Jōkaku kenkyū saizensen. Bekkan rekishi dokuhon*, vol. 71. Shinjinbutsu Ōraishia, 1996.

Sakai Naoyuki, ed. *Shiro no mikata arukikata. Bekkan Rekishi dokuhon*, vol. 103. Shinjinbutsu Ōraishia, 2002.

Shinjō shi shidarakgahara rekishi shiryōkan Nihon saiko no hinawajūten. Shinjō, 2001.

Shinsetsu Hōjō Godai Sōun to ichizoku, hyakunen no kōbō. Gakushū kenkyūsha, 1989.

Shūkan Asahi hyakka. Nihon no rekishi, vol. 21 *Shiro*. Asahi Shinbunsha, 1986.

Shūkan Asahi hyakka. Nihon no rekishi, vol. 24 *Teppō denrai*. Asahi Shinbunsha, 1986.

Shūkan Asahi hyakka. Nihon no rekishi, vol. 98. *Seinan sensō to Ryūkyū shobun*. Asahi Shinbunsha, 1988.

Sugiyama Hiroshi. *Sengoku daimyō*, in *Nihon no rekishi*, vol. 11. Chūkō bunko, 1974.

Suzuki Masaya. *Teppōtai to kiba gundan*. Yōsensha, 2003.

Suzuki Masaya. *Teppō to nihonjin*. Chikuma gakugei bunko, 2000.

Suzuki Tomokichi. *Kyōto fukin chōmei kosenshi*. Kyōto: Matsuzaki publishing, 1936.

Takahashi Masaaki. *Bushi no seiritsu bushizō no sōshutsu*. Tōkyō daigaku shuppan, 1999.

Terashima Yoshiyasu, comp. *Wakan sanzue*, vol. 4. Edited by Shimada Isao, et al. Heibonsha, 1986.

Togashi Yuzuru (ed), *Kassen emaki—Bushi no sekai*. Mainichi shinbunsha, 1990.

Tokubetsuten Kunitomo teppō tanji – sono sekai. Compiled by Shiritsu Nagahamajō Rekishi hakubutsukan, 1985.

Toma Shi'ichi. 'Hiya ni tsuite.' *Nantō Kōko*, no. 14 (December 1994): 123-152

Toyoda Aritsune and Nomura Shin'ichi (eds). *NHK rekishi e no shōtai*. Nihon hōsō shuppankai, 1980.

Udagawa Takehisa. *Teppō to Sengoku kassen*. Yoshikawa kōbunkan, 2002.

Udagawa Takehisa. *Teppō denrai*. Chūkō shinso, 1990.

Yamagishi Sumio et al. (eds). *Nihon kachū no kisoteki chishiki*. Yūzankaku, 1990.

Yamakami Hachirō. *Kachū no shin kenkyū*. 2 vols. Iikura shoten, 1942.

Yamamoto Hirofumi. *Shimazu Yoshihiro no kake*. Chūkō bunko, 2001.

Yamamoto Hirofumi. *Edojō no kyūtei seiji*. Kōdansha bunko, 1996.

BOOKS IN ENGLISH

Arai Hakuseki. *Told Round a Brushwood Fire: The Autobiography of Arai Hakuseki*. Translated by Joyce Ackroyd. Princeton: Princeton University Press, 1980.

Asakawa, Kan'ichi. *The Documents of Iriki*. Tokyo: Japan Society for the Promotion of Science, 1955.

Birt, Michael. 'Warring States: A study of the Go-Hōjō Daimyō and Domain 1491–1590.' Ph.D. dissertation. Princeton University, 1983.

Bottomly, I. and A. P. Hopson. *Arms and Armor of the Samurai: The History of Weaponry in Ancient Japan*. New York: Crescent Books, 1988.

Boxer, C. R. 'Notes on Early European Military Influence in Japan (1543–1853)' in *Transactions of the Asiatic Society of Japan, Second Series*, vol. VIII (1931): 67–93.

Brown, Delmar. 'The Impact of Firearms on Japanese Warfare 1543–98' in The *Far Eastern Quarterly*, vol. 7, no. 3. (May 1948): pp236–253.

Carman, William Y. *A History of Firearms From Earliest Times to 1914*. New York: St. Martin's Press, 1955.

Chase, Kenneth. Firearms: *A Global History to 1700*. New York: Cambridge University Press, 2003.

Conlan, Thomas D. *State of War: The Violent Order of Fourteenth Century Japan*. Ann Arbor: University of Michigan Center for Japanese Studies, 2003.

Conlan, Thomas D. *In Little Need of Divine Intervention: Takezaki Suenaga's Scrolls of the Mongol Invasions of Japan*. Ithaca: Cornell East Asia Series, 2001.

Conlan, Thomas D. 'Instruments of Change: Organizational Technology and the Consolidation of Regional Power in Japan, 1333–1600.' Unpublished manuscript.

Cooper, Michael. *They Came to Japan: An Anthology of European Reports on Japan, 1543–1600*. Berkeley: University of California Press, 1965.

Delbrück, Hans. *The Dawn of Modern Warfare: History of the Art of War*, vol. 4. Translated by Walter J. Renfroe, Jr. Omaha: Bison Books, 1990.

Elison, George and Bardwell Smith (eds). *Warlords, Artists & Commoners: Japan in the Sixteenth Century*. Honolulu: The University Press of Hawaii, 1981.

Fisher, David Hackett. *Paul Revere's Ride*. New York: Oxford University Press, 1994.

Friday, Karl. *Samurai, Warfare and the State in Early Medieval Japan*. New York: Routledge, 2004.

Fukuzawa, Yukichi. *The Autobiography of Yukichi Fukuzawa*. Translated by Ei'ichi Kiyooka. New York: Columbia University Press, 1966.

Hurst, G. Cameron. *Armed Martial Arts of Japan*. New Haven: Yale University Press, 1998.

Lamers, Jeroen. *Japonius Tyrannus: The Japanese Warlord Oda Nobunaga Reconsidered*. Leiden: Hotei Publishing, 2000.

Laws of the Muromachi Bakufu: Kemmu Shikimoku (1336) and Muromachi Bakufu Tsuikahō. Translated by Kenneth Grossberg and Kanamoto Nobuhisa. Tokyo: Monumenta Nipponica and Sophia University Press, 1981.

Lidin, Olof. *Tanegashima: The Arrival of Europe in Japan.* Copenhagen: NIAS Press, 2002.

Lu, David. *Japan: A Documentary History.* Armonk: M. E. Sharpe, 1997.

Mass, Jeffrey P. *Yoritomo and the Founding of the First Bakufu.* Stanford: Stanford University Press, 1999.

Mass, Jeffrey P. *Lordship and Inheritance in Early Medieval Japan: A Study of the Kamakura Sōryō System.* Stanford: Stanford University Press, 1989.

McClain, James L. 'Castle Towns and Daimyō Authority: Kanazawa in the Years 1583–1630' in *The Journal of Japanese Studies,* vol. 6, no. 2 (Summer 1980): 267–99.

'military technology.' *Encyclopaedia Britannica.* 2007. Encyclopaedia Britannica Online. <http://search.eb.com/eb/article-9110174>.

Needham, Joseph. *Science and Civilization in China,* vol. 5, pt. 7, 'Military Technology; The Gunpowder Epic.' New York: Cambridge University Press, 1986.

Needham, Joseph. *Science and Civilization in China,* vol. 5 pt. 6, 'Military Technology: Missiles and Sieges.' New York: Cambridge University Press, 1994.

Perrin, Noel. *Giving Up the Gun: Japan's Reversion to the Sword, 1543–1879.* Boulder: Shambhala Press, 1980.

Pinto, Mendes. *The Travels of Mendes Pinto.* Edited and translated by Rebecca Catz. Chicago: University of Chicago Press, 1989.

Ravina, Mark. *The Last Samurai: The Life and Battles of Saigō Takamori.* Hoboken: John Wiley and Sons, 2004.

Sakakibara, Kōzan. *The Manufacture of Armour and Helmets in Sixteenth Century Japan.* Rutland: C. E. Tuttle, 1962.

Sansom, George B. *The Western World and Japan.* New York: Alfred A. Knopf, 1973.

Shimizu, Yoshiaki. *Japan: The Shaping of Daimyo Culture 1185–1868.* Washington: National Gallery of Art, 1989.

Smith, Thomas. 'The Introduction of Western Industry to Japan During the Last Years of the Tokugawa Period' in *Harvard Journal of Asiatic Studies,* vol. 11, no. 1/2 (June 1948): 130–52.

Takekoshi, Yosoburo. *The Economic Aspects of the History of the Civilization of Japan,* vol. 1. New York: The Macmillan Company, 1930.

Turnbull, Stephen. *Samurai: The World of the Warrior.* London: Osprey Publishing, 2003.

Turnbull, Stephen. *The Book of the Samurai: The Warrior Class of Japan.* New York: Arco Publishing Company, 1982.

Varley, Paul. 'Oda Nobunaga, Guns and Early Modern Warfare in Japan,' in James Baxter and Joshu Fogel (eds), *Writing Histories in Japan; Texts and Their Transformations from Ancient Times through the Meiji Era,* 105–125. Tokyo: International Research for Japanese Studies, 2007.

Varley, Paul, 'Warfare in Japan 1467–1600,' in Jeremy Black (ed.), *War in the Early Modern World,* 53–86. Boulder: Westview Press, 1999.

Yamamoto, Tsunetomo. *Hagakure: The Book of the Samurai.* Translated by William Scott Wilson. Tokyo: Kōdansha International, 1979.

INDEX

Page numbers in *italics* refer to illustrations.

A
abolition of samurai order 205–6, 208, 209
Adams, Will 195
Akamatsu family 109, 123
Akamatsu Ujinori 73
Akechi Mitsuhide 140, 178, 188
Akenajō Castle 148
Aki no yo no nagamonogatari scrolls 72
Akira Kurosawa 85, 166, 167, 172, 177
akoda bachi helmet 63
Akune cannon *185*
akusō warrior monks 75
  see also warrior monks
Amako family 126, 163
Amano Okisada 151
Amida mountains 82
Anayama Nobukimi 173, 174
anti-foreign sentiment 204
Aonogahara, battle of (1338) 82
Arai Hakuseki 105
archery
  accuracy of *10*, 19, 36
  competitions *33*
  horses, against 56–7
  kyūba no michi 'the way of the bow and arrow' 10, 18–19, 49
  mounted 36, 48–9, *53*, *54*, 55–6
  skirmishers 19, 62, *64*, 100
  tactics 21, 55–6, 72
  training *10*
  see also bows
Arima family 197
Arima Harunobu 186
armour *26–7*
  bullet-proof *162*, 175
  chain mail 40, 119, *122*, *124*, 162
  construction 40–1
  cost of 41, 61, 64
  dōmaru style *40*, 62, 64, 72, *92*, 120
  evolution of 117, 119–20, 122–3, *124*, *152*
  female *33*, 64
  haidate leg protectors 46, *118*
  hara-ate armour 62, 64, *73*
  haramaki armour *31*, 61, 62, *63*, 75
  hatsumuri face guard *28–9*, 44, *57*
  horse armour *33*
  kusazuri leg protectors *39*, *42–3*, 64
  mounted warriors *28–9*, 30, *31*, *36*, 38, *39*, 40–1, 57, *113*
  muna-ita breast plate *62*, 64
  oni damari breastplate *124*
  osode shoulder armour *115*, *202*
  ōtate age sune ate shin guards *63*
  ōyoroi style *39*, 40–1, *49*, *61*, 64, *202*
  paper armour 119, 133
  putting on *42–3*
  shortages of 100
  skirmishers 66
  sode shoulder armour *31*, *39*, 40, 41, *42–3*, 44, 48, *49*, *63*, *87*, 152
  sune ate shin guards *39*, 46, 48, 64, *115*, 122
  suso ita leg protection edging *87*, *92*, 150, 152
  tate eri armoured collar 122
  tōsei gusoku armour 117, *118*, 124, 137, 150
  yokohagi-dō armour *87*, 92
  Yuki no shita do armour *120*, 122–3
  see also helmets
armour makers 122–3, 126
arrowheads 51–2
arrows 51–2, *53*, *54–5*, *57*, *106*
  wounds from 56, 57, 62, 163
artillery
  catapults 147, 182, *183*
  trebuchets 74, 100, 182
  see also cannon; firearms
Asakura family 134
Asakura Norikage 129
Asakura Toshikage 129, 134
ashigaru infantry 74, *87*, *92–3*, 100, 133, 138, 164, 171
Ashikaga shoguns 50, 91–2
  established 77, 78–9, 82–3, *104*
  firearms 153, 154, 158
  fortifications and 74
  insignia *63*
  rule of 107, 108–10, 125, 129, 141
Ashikaga Tadafuyu 90, 105, 106
Ashikaga Tadayoshi 90, 106
Ashikaga Takauji 78, 79, 80, 81, 90, 105–7, 108, 113
Ashikaga Yoshiaki 187
Ashikaga Yoshiakira 107
Ashikaga Yoshiharu 153
Ashikaga Yoshimasa 92, 94, 96, 111, 114
Ashikaga Yoshimi 126
Ashikaga Yoshimitsu 108–10
Ashikaga Yoshimochi 110
Ashikaga Yoshinori 89, 110–11, 136
Ashikaga Yoshitane 111, 124, 125, 126
Ashikaga Yoshiteru 101, 154, 157, 159, 160, 184
Aso Harutoki 107–8, 113, 182
atake bune ship *192*
Azuchi Castle 188, 190

B
Baba Nobuharu 173, 174
bajōzutsu firearm *145*
bamboo
  for bows and arrows 49, 51, 100
  pikes 86
bannermen 20, *31*, 74, *77*, 79
banners *102–3*, *107*, *125*, *136*
  see also flags
barbers 206, *208*
battle axes 66, 72–3, *74*
battle formations *130*
beheading 97, 104, 107
boats *48–9*, *54–5*, 90, *179*, 191
Book of Five Rings 12
bows *106*
  composite bows 49
  firearms, compared with 161–2, 176
  firing *22*, *52*
  length of *8*, 49, 56
  marugi yumi 49
  range of 56, 151
  strength of 52
  see also archery
Budō Geijutsu hiden zue 163, *165*, *176*, *184–5*
bullets 148, 149, 161, 170, 197
Bungo Province 153

C
cannon
  breech-loading 183–5
  Chinese *182*, *183*, 184
  earliest 182–7
  gun carriages *200*, 201
  manufacture *184*, 186–7, 192–3

INDEX

monopolization of supply 182, 193–5, 197–9
muzzle-loading 184, 187
rifling 203
see also firearms; gunsmiths
cannon shells 187
castles
  cannon and 187, 188, 192–3, 208
  construction methods 75–7, 154, 187–8, *189*, *193*
  early 73–4
  firearms and 148–9, 154
  Hōjō family 131
  moats 188, *190*, 191, 193
  mountain castles 74–5
  power and 188
  technology 117
  see also individual castles
catapults 147, 181, 182, *183*
cavalry
  tactics 53–6
  urban warfare 73, 83, 98
  see also horses; mounted warriors
chain mail 40, 119, *122*, *124*, 162
China
  cannon *182*, *183*, 184
  firearms 147–8, 151
  Korea *188*
  Opium Wars 201, 208
  pikes 86
  trade with 124, 160
Christian missionaries 126, *146*, 184
Christianity
  conversion to 153, 154, 197
  prohibition of 197
Chronicle of Ōnin 98
Chōshū 205, 206
Chōsokabe family 191, 192
civil wars 7, 21, 77, 95–6
climate 40
clothing *40–1*
  dressing for battle *42–3*, 152
  hitatare robes *42–3*
  see also armour; footwear; headgear
commanders
  Ashikaga rule 108–11
  battlefield control 134, *177*
  becoming famous 103
  changing attitudes to 112–14
  conservatism 123–6, 135–6
  letter-writing 107–8
  new patterns of command authority 114–16
  problems facing 104–5
  rewarding warriors 105, 106–7
  warlords' strategies 123, 141
communications 131, 134
composite bows 49

D
Daigoji 80, 82
daimyō
  boats 91
  expanding armies 109
  firearms 144–5, *147*, 154
  horse armour 33
  last destroyed 115, 134–5
  Meiji control of 205–6
Danjō Hills 167, 168–9, 171, 172, 173
Date Masamune 140
Dejima Island 182, 197, 200, *201*
Dewa province 35
Documents of Iriki 176
dōmaru style armour *40*, 62, 64, 72, *92*, 120
'Dutch' technological knowledge 25, 199, 201, 204

E
ebira quiver *52*, 53
eboshi headgear *40–1*, 44, 48–9
Edo Castle 128
Endō Ishikawa 90
Enfield rifles 205, 206
England 195, 197
Enryakuji temple 111, 112, 141
  see also Mt Hiei
Etō Shinpei 15, 206
explosives 147

F
family system 20–1
  see also individual families
fans 115, 134
firearms
  accuracy of 161, 163, *165*
  armour against 123, *162*, 175
  bows, compared with 161–2, 176
  earliest 100, *101*, 126, 143–4, 147–9
  manufacture of 154, 157
  monopolization of supply 193–5, 197–9
  range of 151, 161, 172–3
  rifling *155*, 157, 203
  spread of 145, 151, 153–4, 163, 166, 175, 176–9, 208
  tactics 163, 165, 166, 170, 176–7
  wounds from 162–3
  see also cannon; gunsmiths; harquebuses
flags *20*, *31*, 88, *93*, *117*, *134*, *138*
flintlocks 199
footwear
  tabi socks 43
  tsuranuki fur shoes *8*, *40–1*, 45, *202*
  waraji sandals *43*, 45, *64*, *73*, *93*, *115*
fortifications
  cannon and 187, 188, 192–3, 208
  construction methods 75–7, 154, 187–8, *189*, *193*
  early 73–4
  firearms and 148–9, 154, *172–3*
  Hōjō family 131
  mountain castles 74–5
  Nagashino, battle of 168–9, 171–2
  technology 117
  trenches 22, 99
France 86
Frederick Barbarossa 40
Frois, Luis 186, 188
frugality, cult of 25, 129
Fujiwara Kunihira 35
Fujiwara Nobuyori *22*, 28–9
Fujiwara Nobuzane 37
fukikaeshi *39*, *44*, *115*, *118*
Fukuzawa Yukichi 201, *203*
furanki cannon 183–4, *185*, 187, 195
fusetake yumi bow 50, *51*
Fushimi Castle 191

G
gauntlets *47*, *63*
  see also kote gauntlets
gekokujō 126–7, *128*
genin, 'the base' stratum of society 30, 32
Genkō and Kenmu Disturbance (1333–38) 60
Ginkakuji (Silver Pavilion) 111, *114–15*
girths, saddle *37*, 38
gloves 53
Go-Daigo, Emperor 20, 74, 78, 79, 82, 106, 108

219

Go-sannen ekotoba picture scroll *50*, 72–3
gokenin, 'honourable housemen' 8, 20, 29–30, 31, 64, 79, 108
Great Britain 201, 204
Great Wall of China 183
gunchūjō petitions for reward 77
gunjingura, military saddle 37–8
gunpowder 144–5, 157–61, 165, 194
guns *see* cannon; firearms
gunsmiths 157, 186–7, 192–5

H

habaki leg protectors 73
haidate leg protectors 46, *118*
hanzei 'half tax' 90, 107
hara-ate armour 62, 64, 73
haramaki armour 31, 61, 63, 64, 75
harquebuses 142–3, 145, 151, 153, 208
    equipment for *152*, *164*
    matchlocks *157*, *163*, *175*
    rifling *155*, 157
    stages of firing *156*
Haruta of Nara 122
Hatakeyama family 109, 110
Hatakeyama Masanaga 92, 94, 98–9
Hatakeyama Yasaburō 91, 92, 115
Hatakeyama Yoshinari 91, 92, 94, 99
Hatano Kegeuji 57
hatsumuri face guard 28–9, 44, 57
Hayashi Shihei 200
head injuries 45
headgear
    eboshi 40–1, 44, 48–9
    hatsumuri face guard 28–9, 44, 57
    see also helmets
Heiji monogatari emaki (Illustrated Scrolls of the Events of the Heiji Era) 22–3, 36, 57, 58–9
heishi warriors 73
helmets
    akoda bachi 63
    evolution of 120, *121*, 122
    fukikaeshi 39, 44, 115, 118
    head injuries and 45
    hō-ate throat guard 44–5, 123, 150, *162*
    jingasa 92, *171*
    kawari bachi 118
    kuwagata helmet horns 39, 44, *201*
    maidate helmet decoration 40, 63, 113, 115, 123, *140*

me no shita hō face guard 123, *124*, *162*, 202
mounted warriors 39, 40, 44–5, 57
shikoro neck guard 113, *118*, 162
ōboshi hachi 50
uchimayu visor 162
zunari 87, *120*, 134, 150
Hiei, Mt 78–9, 80, 82, *83*, 111, 112
    see also Enryakuji
Higashiyama 75
Himeji Castle 188
Hino gunsmiths 198
Hiraizumi 35
Hirosaki Castle 23
hirosode flexible sleeves 61
Hiryūsō firearms *147*, *148*
Hitachi Province 122
hitatare robes 42–3
hiya 'fire arrows' 149
Hizen Province 197
hō-ate throat guard 44–5, *123*, *150*, 162
Hōgen Disturbance (1156) 6–7
Hōjō family *30*, 116, 117, 128, 137, 139, 141, 165, 178
Hōjō Sōun 126–7, 128, 129, 131, 151
Hōjō Ujimasa 177
Hōjō Utjitsuna 128, 129, 131, 132
Hōjō Ujiyasu 132
Hōjōji 79, 80, 82
Holland 182, 187, 191, 197, 199
Honda Tadakatsu *122*, 135, 137
Honda Tadatomo *135*
Honganji Temple 188
horses
    armour 33
    character of Japanese 32–3, 35
    Mongol 35, 36
    speed of 35–6
    tack 37–8, *63*
    valued by Japanese 30, 32
    wounds to 56–7, 69, 71, 89
    see also cavalry; mounted warriors
horseshoes 35
hosammai kitae sword blades 66
hoshi-kabuto helmet *121*
Hosokawa family 91, 96, 97, 109, 110, 123, 126, 127
Hosokawa Fujitaka 170
Hosokawa Harumoto 101, 153
Hosokawa Katsumoto 94–5, 100, 127
Hosokawa Masamoto 111, 123, 124–5, 127–8
Hosokawa Sumitomo 63

Hosokawa Yoriyuki 108
Hō Amidabutsu 30
Hōkōji bell *196*
hyakushō, hundred names 32

I

Ichi-no-tani, battle of (1184) 36, *49*, 98–9
Ichijō Nobutatsu 173
igi, saddle boards 37, *38*
Ii Naomasa 178
Imagawa family 123, 126, 127, 128–9, 136, 179
Imagawa Ryōshun 108
Imagawa Ujichika 127, 128
Imagawa Yoshimoto 116, 129
Imperial Japanese Army 18
Inadome family 187
Inami warriors 132
inheritance, land 30
Inland Sea 97
insignia *20*, 117, *120*, *125*, 128, *150*, *162*, 194
Ippen Hijiri e scrolls 76
Ise Province 122
Ishiyama Honganji 179
Ishiyama Temple 159, *160*
Isshiki Yoshitada 95
Itakura Shigemasa 197
Izumi Province 100, 163, 182

J

jingasa helmet *92*, *171*
Jingū kōgō engi emaki 89, 90
jitō, land stewards *28*, 30, 31
Jōei formulary 29
jumonji pike 89

K

kaburaya humming arrow *52*, 54
kabuto helmet *121*
kagi yari pike 89
Kagoshima 204, *205*, 206–7
Kai Province 138, *145*
Kaizu Castle 117, 139, 187
kama yari pike 89
Kamakura *bakufu*
    destruction of 20, 31, 78, 108, 113
    judicial activities of 29–30, *48*
    Latter Hōjō and 127
    Mongol invasions 104
    restrictions on horse riding 30
Kamishiro Kaneharu 79, 82

INDEX

Kanazawa Castle 188, *190*, 191
Kanezawa Sadaaki 40
Kannonji Castle 187
kanrei deputy shogun 108, 109, 110, 111, 127, 135–6
kantō kanrei 135–6, *139*
karimata arrowhead *51*, *52*
kassen chūmon battle reports 77
Kasuga Castle 177, 187
Kasuga gongen kenki e picture scroll 64, 72
katakami armour cords 61
katana sword 8, *15*, *19*, 67, *72*, *75*, 145
  drawing 65, 66
  growing popularity of 68
  guard posture 65
  hand grip 69
  types of stroke *70*, *71*
Katayama Takachika 79
Katsurenjō Castle 148, *149*
Kawachi Province 91, 92
Kawanakajima, battle of (1561) 134, 138–41, 165, 175
kawari bachi helmet *118*
kendō 12–14
kenjutsu 13
Kii Province 91, 92, 95
Kikuchi pike 86
Kikuchi Takefusa 54, 104
Kinkakuji (Golden Pavilion) 109, *111*
kinpaku oshi juzu rosary *122*
Kitabatake Akiie 82, 104
kiyumi bow 49, *51*
knives *15*, 67, 68, *68–9*, 145
Kō no Moroakira 46, 53
Kō no Moronao 79, 80, 83
Kobayakawa family 179
Kobayakawa Takakage 161
kokujin, 'men of the province' 31
Komaki Nagakute campaign 177–8
Konishi family 197
Konoe Sakihisa 139
Konoe Tane'ie 158–9
Korea 20, 124, 150, 176, 182, 188, 191
kote gauntlets 40, *42–3*, *46*, *47*, *48*, 64, *118*, *162*, *202*
kozuka knife 68–9
kukishin ryū kenpō swordsmanship *16*
kumade hooks 55
Kumamoto Castle 207
Kunitomo gunsmiths 157, 186–7, 193, 194, 208

kurabone, saddle seat 37
kusari bakama chain mail leggings *124*
kusazuri leg protectors 39, *42–3*, 64
Kusunoki Masashige 78, 113, 129, 182
kuwagata helmet horns 39, 44, *201*
Kyoto 94–5
  battle for (1336) 78–83, *104*, 105–6, 153–4
  destruction of 92, 96, 100
  Ōnin War 96–7, 98–100, 123, 148
  urban warfare 73
kyūba no michi 'the way of the bow and arrow' 10, 18–19, 49
Kyushu 40, 76

L

land
  disputes over 29
  inheritance 30
  jitō, land stewards *28*, *30*, 31
  military service, rewarded for 105, 132
  redistribution 7, 131
  warriors' links with 8, 10–11
Last Samurai, The (film) 15
Latter Hōjō family 116, 117, 119–20, 123, 126–9, 131–5, 177
legal system
  Hōjō Sōun 127
  Jōei formulary 29
letter-writing 107–8
logistics 22, 23, 83, 86, 90–1, *94*

M

Macao, cannon *186*, 187, 197, 199
Maeda Toshinaga *176*
maewa, pommel board 37–8, *53*
maidate helmet decoration 40, *63*, *113*, *115*, *123*, *140*
makuri kitae sword blades 66
martial arts 11–12, *13–14*, 15
maru kitae sword blades 66
marugi yumi bows 49
masakari battle axe 66, *74*
Masuda Muneharu 39
matchlocks *157*, 163, 175
Matsumoto Castle 188
Matsura family 154
me no shita hō face guard *123*, *124*, *162*, *202*
Meiji era *18–19*, 205, 206, 207
menuki sword decorations 67, *68*

Migita family 20
Mimigawa, battle of (1578) 186
Minamoto Yoritomo *21*, 28–9, 33, 105
Minamoto Yoshiie *50*
Minamoto Yoshitomo 28–9, *57*
Minamoto Yoshitsune *36*, 49
mitrailleuse 86
Miura family 127
Miyagi Shirō no jō 133
Miyamoto Musashi 12, *17*
Miyoshi Nagayoshi 101
moats 188, *189*, 191, 193
mobilization 116, 131, 132–4
mon symbols *117*, *138*
Mongols
  firearms 147
  horses *35*, *36*
  invasions of Japan *30*, *31*, 48–9, 54, *55*, 76, 77, 90, 104
  ships *90*, *104*
  see also Scrolls of the Mongol Invasions
monks *75*, 111, 112, *152*
momonari-kabuto helmet *121*
Mōri family 185, 195
Mōri Katsunaga 135
Mōri Motonari 160
mounted warriors
  armour 28–9, *30*, *31*, *36*, 38, *39*, 40–1, 57, *113*
  hyakushō, hundred names 32
  importance of 30–1
  tactics 53–6, *57*, 100, 170–1
  see also cavalry
muna-ita breast plate *62*, 64
munegai, horse's chest strap *37*, 38
Murakami Yoshikiyo 138
Musashi Province 88, 89, 128
muskets 151, 153, 161, *175*, 195, 199
Mutsu province 35
Myōchin Nobuie 120, *122*
Myōshu 31

N

Nagahama 187, 193, *194*
Nagao Kagetora 136
  see also Uesugi Kenshin
Nagasaki 182, 197, 199, 200, 201
Nagashino, battle of (1575) 24, 129, 165–75, 177, 178
naginata glaive 11, 38, 66, *72*, 73, *75*, 86
Nagoya 191
Naitō Masatoyo 171

Naitō Tadatoshi of Tanba 150
Nakagusukujō Castle 148
Nakazawa family 91
Nanban byōbu screens *146*
Nanpō Bunshi 144
Nativism 204
Nawa Nagatoshi 79, 80, 82
Negoroji Temple 143, 152, 154, 157, 163, 179
Nijō Castle 187–8
Nirayama Castle 117, 127
Nitta Yoshisada 78, 79, 80, *81*, 82, *104*, 106, 113, 129
Ni'u Castle 185
nobori flags *138*
noboru flag *93*
nobushi skirmishers 62, *64*
Nochikagami 153
nodowa throat protector 39, 40, *42–3*, 63, *202*
Nomi Motonobu 161
'Northern White Flag Corps' 88
nō plays 111
Nyoigadake Castle 153–4

O

ō-bashi kabuto helmet *121*
Oda Nobunaga 87, 103, 109, 112, 114, 116, 133, 141, 145, 152, 174
  death 178, 188, 191
  fortifications 187–8
  Ishiyama Honganji 179
  Komaki Nagakute campaign 177–8
  Nagashino, battle of 166, 167, 169–70, 171, 173, 174–5
Ōdachi Harumitsu 154, 159
ōdachi long sword 63, 68, 71, 72
Odawara Castle 127, 134, 142–3, 165
Ogasawara Nagatoki 138
Ogasawara Sadamune 81
Ogasawara Ujihira 80
Okamoto Hachirō Saemon Masahide 132–3, 135
Okinawa 148, 151
  see also Ryūkyū Kingdom
okisode sleeves *118*, *150, 162*
Ōkubo Tadasuke 172
Ōkubo Tadayo 172
Ōmi province 117
Ōmori Ujiyori 127
oni damari breastplate *124*
onigiri rice balls *93, 96–7*

Ōnin War (1467-77) 22, 60, 62, 92–101, 103–4, 111, 115, 123–4, 127–8, 148
Opium Wars, China 201, 208
organization, military 83, 86, 88–90, *130*, 131–5, 137–8, 165
Osaka Castle *91, 133*, 135, *180–1*, 188, 189, 191, 193, *194–5*, 199
Ōshio Heihachirō 199, 200
ōsode shoulder armour *115*, 202
ōtate age sune ate shin guards *63*
Ōtomo family 163, 185–6
Ōtomo Yoshishige *153*, 154, 159, 184, 186
Ōuchi family 90, 110, 123–6, 128, 129, 131, 138, 141
Ōuchi Masahiro 97, 123, 125
Ōuchi Yoshioki 125–6
Ōuchi Yoshitaka 126, 154
Owari Province 123
Ōyama zumi shrine 64, 88
ōyoroi style armour 39, 40–1, *49, 61, 64, 202*

P

paper armour 119, 133
Perry, Commodore Matthew 201, 203, 204, 205, 209
Philippines *19*, 150
pikemen
  armour *87, 92, 93*
  equipment *93*
  Ōnin War *95, 96*, 97, *99*, 124
  training 22, 86, *88, 95, 109*
pikes 11, *84–5*
  horses, against 56, 57, 86, 87
  importance of 86, 88, 132, 178, 179
  length of 87, 92, *109*, 133, 179
  skirmishers 66, 72, *195*
  swords, against 97
  tactics 86, *91, 96*, 124, 130
  types of *89*
  wounds from 56, 57, 88, 97
Pinto, Mendes 144, 158
Portugal *146*
  cannon 181, 184, 186, 187
  firearms 23, 101, 126, 143, *152*, 153, 154, 163, 181
  ships *144*, 192

Q

quivers *52, 53, 106*

R

Ran (film) *84–5, 177*
Record of the Musket 144, 147
reins 38, *53*
Rengo River 167, 168, 169
rewards for fighting 20, 77, 78, 89–90, 104, 105, 106–7
rokkaku antlers *122, 135, 137*
Rokkaku Yoshiharu 187
Rurikōji Temple 124, 125
Ryōunji Temple 125, *127*
Ryūkyū Kingdom 148, 201
  see also Okinawa
Ryūzōji Takanobu 186

S

saddles 37–8
Saga Province 25, 201
Sagami Province 127
Saigō Takamori 15, *18–19*, 25, 206–7, *208–9*
Saijosan 139, 141
Saitō Toshimitsu *140*
Sakai *154–5*, 157, 187, 192, 193, 194
Sakuma Nobumori 173–4
Sakuma Shōzan 25
saltpetre 157, 159–60
samurai (followers) 8, 31, 32, 64
Sanbōin temple 97
sane armour plates 40, 41, 44, *61*
sanmai uchi yumi bow 49, *51*
sashimono banners *107, 117, 136*
Satomi family 131
Satsuma 201, 202, 203, 204, 205
scabbards *15, 67, 68–9*
Scrolls of the Mongol Invasions 20, *30, 40–1, 44, 45, 53, 77*, 86, 90, 147
Seki castle 74, 75
Sekigahara, battle of (1600) 129, 135, 150, 168, 178, *179*, 191
The Shadow Warrior (film) *166–7*
Shiba family 109, 110
Shibatsuji Ri'uemon 192
shields 64, 66, 72, 96, 100, 195
shihō-zume kitae sword blades 66
shihōchiku yumi bow *51*
shikoro neck guard 113, *118, 162*
Shimabara 197, *198–9*
Shimazu family 123, 153, 178, 185–6, 197, 205
Shimazu Hisachika 48–9
Shimazu Iehisa 186
Shimazu Nariakira 201, 202, 203

Shimazu Yoshihisa 176
Shinano Province 138, 141
ships
    British 204
    cannon on 185
    Dutch 195, 197, *200*
    innovations in *192*
    Korean 191
    Mongol *90*, 104
    Portuguese *144*, 192
shirigai, horse's rear strap *37*, 38
shirushi identification symbol *113*
shitagura, under-saddle 37
Shizugatake, battle of (1583) 8
shizuwa, saddle cantle 37–8
Shōkokuji Temple 98, 99, 109
Shōni Tsunesuke *48–9*
Shōni Yorinao 79, 82
Shōrakuji Castle 154
shugo 'protectors' 23, 90–1, 92, 95, 101, 109, 110, 116, 127
sidearms 145
sieges 74–5, *133*, 188
    tactics *190*, *195*, *197*, 198
skirmishers 62
    archery 19, 62, *64*, 100
    armour 66
    tactics 62, 64, 66, 71–2, 73, 100, *195*
sode shoulder armour 31, *39*, *40*, 41, *42–3*, 44, 48, *49*, *63*, *87*, *152*
    see also ōsode
'Southern Corps' 88
Southern Court 74, 80, 82
standard bearers *117*
steel armour 123
stirrups 38
Sue Takafusa 126
sune ate shin guards *39*, *46*, 48, *64*, *115*, *122*
suso ita leg protection edging *87*, *92*, *150*, *152*
swords
    blade distinctions *66*
    construction 67, 157
    decline in use of 88
    fittings 67–8
    importance of 18, 66–7, 88, 208
    lengths 68, 71, 72
    ōdachi long sword *63*, 68, 71, 72
    sharpening *60*
    skirmishers 66, 71–2
    tachi sword 65, 67, *68–9*, *72*
    training 12, *13–14*, 15, *16–17*

wakizashi sword *68–9*, *72*, *87*, *92*, *106*, *145*
wounds from 56, 57, 69, 71, 88, 97
see also katana and *tachi* swords

T
tabi socks 43
tachi sword 65, 67, *68–9*, *72*
tack 37–8, *63*
tactics
    archery 21, 55–6, 72
    battle formations *130*
    evolution of 22, 89, 100–1
    firearms 163, 165, 166, 170, 176–7
    mounted warriors 53–6, 57, 100, 170–1
    pikes 86, *91*, *96*, 124, 130
    siege 75, *133*, *190*, *195*, *197*, 198
    skirmishers 62, 64, 66, 71–2, 73, 100, *195*
Tadarahama, battle of (1336) 78
Tahara Naosada 82
Taiheiki (Record of a Great Pacification) chronicle 113, 182
taihō cannon *187*
Taihō castle 74, 75
Takashima Shunhan 199, 201, 205
Takeda family 102–3, 123, 128–9, 136, 140, 141, 145, 151, 165
    destroyed 178
    Nagashino, battle of 166, 167, 168–9, 170–1, 172–5, 177
Takeda Katsuyori 166–7, 174, *175*, 176
Takeda Nobutora 129
Takeda Shingen 103, 113, 114, 117, *128*, 129, 134, 141, 187
Takezaki Suenaga 20, *31*, *46*, 54, 104
Tale of the Heike 35
Tanegashima 143, 148, 154, 158, 163
Tanegashima family 153
Tanegashima Hisatoki 144
Tanegashima Tokiaki 158–9
Tanegashima Tokitada 143
tantō knives 15, *68*, *145*
tanzutsu firearm *145*
Tarpan horse 32, 33
tate eri armoured collar *122*
taxation 90, *94*, 107, 109, 116
teari porters *106*
technological innovations 25, 100–1, 124, 199, 201, 203, 208–9
Tendai Buddhism 110, 111, 141

teppō firearms 143, 144, 147, 151, 153
Togashi Taka'ie 81
tōjin gashira helmet *121*
Tōji Temple 78, 79, 105–6
Tokugawa Iemochi 158
Tokugawa Ieyasu *122*, *129*, 168
    cannon 191–5
    Nagashino, battle of 167, 170, 171, 174
    Osaka Castle 189, 191, 193, 197
Tokugawa shoguns 12, 39, 105, 128, 178
    cannon 25, 182, 197, 208–9
    fall of 24, 205, 209
    knowledge, restrictions of 25, 199, 200, 201, 203, 209
    Nagashino, battle of 168–9, 171, 173
Tokugawa Yoshimune 199
Tokugawa Yoshinobu 39, 205, *206*
Tomoe Gozen *34*, 35
tori-kabuto helmet *121*
tōsei gusoku armour 117, *118*, *124*, *137*, *150*
Toyoshima Iehide 82
Toyotomi Hideyori 189, 191, 193, 195, 197
Toyotomi Hideyoshi 9, 129, 134, 152, 178, 179, 189, 191, 193, 195
    land and 8, 10
tozama, 'outsiders' 31–2
training
    archery *10*
    pikemen 22, 86, *88*, *95*, *109*
    swords 12, *13–14*, 15, *16–17*
trebuchets 74, 100, 182
trenches 22, 99
tsuba sword guards *15*, *67*
Tsuchimochi Nobuhide 7, 8, 18
Tsuchiya Masatsugu 173, 174
tsuka ito 15
tsuranuki fur shoes *8*, *40–1*, *45*, *202*
Tsurugaoka Hachiman 128, *132*
tsuruwa bowstring container *8*, 53
Tsutsui family 91

U
uchimayu visor *162*
Uchino 79
Uesugi family 116–17, 123, 127–8, 129, 134, 135–41, 163, 165, 198, *203*
Uesugi Kenshin 103, 113, 114, 124, 131, 136–41, 154, 160, 175

see also Nagao Kagetora
Uesugi Seishi 135
Unkai Mitsunao 162
Unsei Daigoku 148
urban warfare 73, 83, 98
Utsunomiya Shigefusa 10

V
Victoria, Queen 39

W
waidate armour section 61, 62
wakizashi sword 8, *68–9, 72, 87, 92, 106, 145*
'war cries' 54
waraji sandals *43, 45, 64, 73, 93, 115*
wariha kitae sword blades 66
'Warring States' era 111, 114, 115, 116, 126, 135, 166
warrior monks *75*, 152
see also akusō
Watanabe Sūkaku 194–5

watchtowers 22, 79, 99
Watonai Sankan *161*
wheellocks 151, 153
women warriors 30, *34*, 35, 64, 119
wounds 56–7, 60, 62, 69, 71, 72, 88, 97, 151, 162–3

X
Xavier, Frances 154

Y
yabusame archery *33*
yagura watchtowers 79
Yagyū Muneyoshi 12
yahoro quiver 52, 53
Yamagata Masakage 171–2, 174
Yamaguchi 123–4, 125
Yamakami Hachirō 120
Yamamoto Tsunetomo 7, 8, 11, 18
Yamana family 96, 107, 109, 110, 123
Yamashima Tokitsugu 153
Yamato, Tsutsui of 91

yari pike *11*, 66, 72
see also pikemen; pikes
yokohagi-dō armour *87, 92*
Yokose Utanosuke Narishige 154, 157
Yonezawa Yamagata 203
Yoshino 74
yugake gloves 53
Yūki kassen emaki picture scrolls 53, 89
Yuki no shita dō armour *120*, 122–3
yurugi no ito 162

Z
zunari helmet *87, 120, 134, 150*

Picture Credits and Acknowledgements

All black-and-white line artworks produced by Wes Brown.
All maps © Amber Books Ltd.

AKG-Images: 49r, 61, 88, 166/167t (Toho/Album)
Art Archive: 10 (Kitano Temmangu Kyoto), 34, 37 & 48/49 (Laurie Platt Winfrey), 50 & 55b (Laurie Platt Winfrey), 60 (Okura Shukokan Museum), 68bl (Oriental Art Museum, Genoa), 76t (Laurie Platt Winfrey), 76b (Tokyo National Museum), 78 (Victoria & Albert Museum), 79 (Gianni Dagli Orti), 90 (Laurie Platt Winfrey), 94t (Tokyo University), 96/97 (Gianni Dagli Orti), 105 (Tokyo University), 122, 126 (Gianni Dagli Orti), 144 (Suntory Museum of Art), 151, 153 (Daitokuji Temple), 201, 204/205
Board of Trustees of the Armouries: 8, 15, 25, 38, 44, 52 both, 67, 68/69t, 68/69br, 89 both, 120 both, 147, 150, 155c&b, 163, 165, 176, 184/185
Bridgeman Art Library: 6/7 (Boltin Picture Library), 9 (Ashmolean Museum), 22/23 (Fitzwilliam Museum), 47 (Leeds Museums & Art Galleries), 83t (Maidstone Museum & Art Gallery), 94/95 & 98/99 (Tokyo Fuji Art Museum), 104, 119 (Heini Schneebeli), 123t & 124 (Leeds Museums & Art Galleries), 145 (Bonhams), 146 (Giraudon), 160b Ashmolean Museum), 200 (Archives Charmet)
Thomas D. Conlan: 30, 31, 34, 40/41, 46, 48b, 53, 54/55, 56, 74, 77, 82, 114/115, 125, 127, 149b, 167b, 168/169tl, 169tr&b, 186/187b, 205b
Corbis: 12 & 18/19 (Asian Art & Archaeology), 19r (Bettmann), 21 & 24 (Asian Art & Archaeology), 29r (Burstein Collection), 33b (Kimimasa Mayama), 35 (Barry Lewis), 36 (Burstein Collection), 45 (Asian Art & Archaeology), 58/59 (Burstein Collection), 63 (Sakamoto Photo Research Laboratory), 80 (Asian Art & Archaeology), 84/85 (Gérard Rancinan), 101 (Asian Art & Archaeology), 107 (Burstein Collection), 111 (Brooks Kraft), 112 (Christophe Boisvieux), 132 (Michael S. Yamashita), 136 (Asian Art & Archaeology), 137 (Charles & Josette Lenars), 138/139 & 140 (Asian Art & Archaeology), 142/143 (Earl & Nazima Kowall), 158/159 (Asian Art & Archaeology), 161 (Peter Harholdt), 168l (Sakamoto Photo Research Laboratory), 177 (Gérard Rancinan), 191 (Michael S. Yamashita), 206/207 (Asian Art & Archaeology)
Dorling Kindersley: 73, 118
Getty Images: 11, 23r, 26/27, 83b (Hideaki Tanaka), 123b, 131 (Noriko Yamaguchi), 149t (DAJ), 154b, 160t & 175 (AFP), 178/179t (Takaaki Motohashi), 183 (John Warden), 186t (Time Life Pictures), 189t (Hiroaki Otsubo/Sebun Photo), 202
Heritage Image Partnership: 193 (British Library)
Mary Evans Picture Library: 32, 33t
Photolibrary: 102/103, 116, 128, 154/155t, 174, 189b, 190, 198/199, 203b
Public Domain: 86, 110 both, 178b, 188, 196, 203t, 205t, 206l, 208/209
TopFoto: 39 (Ancient Art & Architecture Collection), 172, 208l (Alinari Archive)
Trustees of the British Museum: 162
Werner Forman Archive: 28/29 (Boston Museum of Fine Arts), 40l (Victoria & Albert Museum), 57 (Boston Museum of Fine Arts), 91 & 133 (Kuroda Collection), 134 (National Museum, Kyoto), 135 & 180/181 (Kuroda Collection), 194t&b (Kuroda Collection)